crafting by concepts

crafting by concepts

fiber arts and mathematics

edited by

SARAH-MARIE BELCASTRO

CAROLYN YACKEL

CRC Press
Taylor & Francis Group
Boca Raton London New York

CRC Press is an imprint of the
Taylor & Francis Group, an **informa** business

AN A K PETERS BOOK

CRC Press
Taylor & Francis Group
6000 Broken Sound Parkway NW, Suite 300
Boca Raton, FL 33487-2742

First issued in hardback 2019

© 2011 by Taylor & Francis Group, LLC
CRC Press is an imprint of Taylor & Francis Group, an Informa business

No claim to original U.S. Government works

ISBN 13: 978-1-56881-435-3 (hbk)

**Visit the Taylor & Francis Web site at
http://www.taylorandfrancis.com**

**and the CRC Press Web site at
http://www.crcpress.com**

Library of Congress Cataloging-in-Publication Data

Crafting by concepts : fiber arts and mathematics / edited by Sarah-Marie Belcastro, Carolyn Yackel.
   p. cm.
Includes bibliographical references and index.
ISBN 978-1-56881-435-3 (alk. paper)
 1. Needlework–Patterns. 2. Needlework–Mathematics. I. Belcastro, Sarah-Marie. II. Yackel, Carolyn.

TT653.C73 2011
746.4–dc22

2010039068

For my parents, Erna and Jim Yackel, who encouraged me in all of my creative pursuits, including mathematics and including fiber arts, and who instilled in me to view every endeavor intellectually and my life holistically so that my work in mathematical fiber arts was made possible.

—Carolyn Yackel

For the three friends, whose ideas inspired my knitting work during the editing of this book, and to whom I turn first when I need my mathematical knitting thoughts and questions to be instantly understood: Rachel Shorey, Amy Szczepański, and Carolyn Yackel.

—sarah-marie belcastro

We would also like to remember our close pets Allie (Carolyn's dog) and Arzachel (sarah-marie's cat), who were our steadfast companions through *Making Mathematics with Needlework* but neither of whom made it to the completion of this volume.

CONTENTS

ACKNOWLEDGMENTS

Our thanks are many. We are grateful first and foremost to the chapter authors for making this book a reality. In this same vein, special thanks go to Alice Peters for her continued confidence in our work, and to Charlotte Henderson for her editorial work on this volume and on *Making Mathematics with Needlework*. (By the way, Charlotte does amazing work on other A K Peters titles, too.) We are also thankful that Apple Computer added screen sharing to iChat, as that made it possible for us to edit this book together while 1,000 miles apart.

Some of the work showcased in this book was first presented at an American Mathematical Society (AMS) Special Session in Mathematics and Mathematics Education in Fiber Arts held at the January 2009 Joint Mathematics Meetings in Washington, DC. We are grateful to the AMS for approving and hosting our Special Session and accompanying Mathematical Fiber Arts Exhibit. Thanks go to Courtney Gibbons for letting us reprint her Brown Sharpie comic (and for making it in the first place), and to Chaim Goodman-Strauss for graciously supplying 14 finite spherical symmetry group diagrams.

sarah-marie thanks her 2009-10 Calc II students for trying out the mathematics in Amy Szczepański's chapter, her parents for giving her the awesome camera (Pentax Optio S12) that took many of the photographs in this book and for being sounding boards in general, and Sean Kinlin for helping with photography, meals, and last-minute everything. sarah-marie is also grateful to her face-to-face crafting buddies (Chelsea, Jasmina, Jenn, Juliet, Miriam, and Rachel) for tolerating her ravings about the book and supplying many forms of feedback and assistance.

Carolyn thanks Creighton Rosental for his breadth and depth of support and dogs Allie and Zeke for their unreserved loving affection and sense of humor. Carolyn would also like to thank George S. Coke of Coke's Camera Shop in Macon, Georgia, for his (repeated) patient instruction and assistance with her camera that allowed her to photograph temari balls.

We both are appreciative of Tom Hull for mathematical, technical, and artistic consultations (individual and joint). We also thank Jess, Casey, and Bob Forbes for making ravelry.com such a central meeting place for all crafters, not just the knitters and crocheters for whom the site is intended, so that we were able to use ravelry to get weaving questions answered (thanks, Nancy Alegria!) and to locate temari and tatting testers quickly and easily.

Speaking of project testers, we are thankful that Brent Annable, Joan Gallant Belcastro, Jennifer Belden, Jasmina Chuck, Margaret Eilrich, Sharon Frechette, Charlotte Henderson, Leslie Hodges, Angela Juliet Larke, Zia Marek-Loftus, Hope McIlwain, Katharine Merow, Kelly Meyer, John Nance, Jeri Riggs, Miriam Roberts, Susan Schmoyer, Lori Schreiber, Ida Gallant Thornton, and Dan Zook tested projects for this book. Some testers were trying these crafts for the first time, and we applaud their efforts. Others are professional-level crafters, and we are especially grateful for their participation. Many of the authors tested projects as well. (See the Credits listing for details.)

Thanks also to our patient photo-shoot models Jenn, Jonah, and Josh Belden, Tom Hull, Sean Kinlin, Chelsea Land, Rachel Shorey, and Veronika Šifrar, and to amateur photographers Heidi Ashton, Mark Eilrich, Marian Goldstine, Neil Hatfield, Marko Šifrar, Steve Swanson, and Laurie Wall for lending us their talents.

Since the publication of *Making Mathematics with Needlework,* we have discovered more individuals who enrich the community of fiber artists who intersect with mathematicians. These include

⋆ Miriam Quinn of the *Knit Science* podcast,

⋆ Pat Ashforth, Steve Plummer, and Ben Ashforth of *Wooly Thoughts* in the UK,

⋆ Brent Annable, Ariel Barton, Hugh Griffiths, Jeny Carden Staiman, and Matthew Wright, for knitting and crocheting cool new mathy things, and

★ the approximately 60 new members of our mathematical fiber arts email list with whom we've connected at national mathematics conferences.

From Chapter Authors. Ted Ashton first thanks his wife for her various supplies, plentiful support and good advice. He also thanks his daughter for providing thread for the cutwork and string art triangles and Susan for teaching him how to tat in the first place. Irena Swanson thanks the testers for very helpful feedback. She also thanks Ellen Kirigin and her mother for introducing her to quilting in her first year in the US. Diane Herrmann thanks Ann Strite-Kurz for helpful conversations about diaper patterns, and EGA and NAN members for their support of her work in the needle arts. For mathematical insights, she thanks Ariel Barton, Thomas Church, and Vipul Naik. Susan Goldstine thanks St. Mary's College of Maryland for nurturing mathematical outreach, and the wonderful staff at Crazy for Ewe in Leonardtown, Maryland, for encouraging her burgeoning passion for knitting. sarah-marie belcastro thanks Tamara Veenstra for saving her by solving two modular equations and Carolyn Yackel for asking the question that started her yarn entanglement investigations.

INTRODUCTION

welcome back

SARAH-MARIE BELCASTRO AND CAROLYN YACKEL

1 Prologue

When we began the project that became *Making Mathematics with Needlework* [1], we had no idea that material for a second book would be generated so quickly, or even that there would be enough interest to merit a second book. Yet, here we are introducing *Crafting by Concepts*, and as we send this volume to press, we are already aware of several works in progress for contribution to a third book.

The Introduction to [1] functioned as a survey of the field of mathematical work on the fiber arts. We will not cover that ground again here. Instead, we describe this book's structure in Section 2, mention recent research themes in Section 3, give comments to educators in Section 4, indicate how to use this book in Section 5, and conclude with Section 6.

2 Chapter Structure

Each of the Chapters and Minis in this book came into being because the authors, who are simultaneously mathematicians and crafters, noticed the potential for the expression of a mathematical idea within certain craft media.

Every Chapter begins with an Overview section in which the interplay between a mathematical notion and the craft expression of this notion is presented for a lay audience. This is followed by a Mathematics section, wherein the ideas are discussed via the formal language of mathematics and the interplay is made explicit. Next, a Teaching Ideas section contains suggestions for fiber-arts-related classroom activities that introduce or reinforce the mathematical concepts discussed in the Mathematics section. In addition, most chapters suggest extensions to the main material in the form of project ideas or open-ended problems. Each Crafting section contains projects that are specifically designed to illustrate

the mathematics in the chapter. Often mathematics is conveyed through the crafting technique, though for the initiated mathematician the result is usually apparent in the finished object.

The two Minis are shorter pieces (hence the name). Each summarizes the theory behind a mathematically inspired project and then moves directly to the Crafting section.

3 Research Update

In [1, pp. 4–7], we outlined four categories of mathematics research in the fiber arts. Since that time, the most work has been done in illustrating mathematical concepts using fiber arts. Anecdotal evidence [3] suggests that it is becoming more popular to answer problems arising in fiber arts by using mathematics. There has also been progress in determining which mathematical concepts can (and cannot) be constructed using a given fiber art. However, little work seems to have been done in describing the mathematics that is intrinsically present in a given fiber art.

We encourage readers to make their own contributions to the literature, formally or informally. We attempt to be cognizant of all publications that combine mathematics and fiber arts, and share the current state of our knowledge via the reference lists given on pages 243–245. These have been updated from [1, pp. 7–10].

4 A Note for Educators

In the Teaching Ideas sections of this book, the authors suggest instructional implementations of the concepts introduced in the Mathematics sections. The goal is to supply hands-on investigation of each mathematical construction. However, these are intentionally not given in the form of worksheets or lesson plans, though the investigations can and should be turned into classroom materials by individual instructors. The reason that the authors have avoided specificity is that classrooms are too varied for such proscription to be useful. The grade level (e.g., secondary, graduate), pedagogy used (e.g., active learning, lecture), and even the context (e.g., homeschool, public institution, math club) all contribute to make each classroom highly individual.

Moreover, we see teaching and learning as highly personal endeavors, taking place in the context of relationships formed between teachers and students in individual classrooms. As such, we have provided the teaching ideas sections as resources from which a teacher can make use of what is most beneficial for her students, elaborating upon it and reworking it so as to fit seamlessly into her course. We hope the following text will assist educators in customizing the material supplied in this book and framing it for their classrooms.

Teaching students mathematics through the approach of crafting is both indirect and subtle. The mathematics that a mathematician sees may not be readily apparent to a student not yet familiar with that mathematical content. In general, we see what we already understand, not the other way around. Therefore, in order for students to learn mathematics through engaging in the fiber arts, it is incumbent upon the teacher to help students come to understand the relevant mathematical aspects of the kinesthetic activity in which they are engaged. In other words, students will not learn mathematics simply by knitting an object. Rather, it is through the accompanying discussion that the relationship between the mathematics and the fiber art activity begins to take on significance for the student and thereby has the potential to deepen the student's understanding of the relevant mathematical content.

Mathematicians are trained to notice numeric, geometric, and abstract relationships. Similarly, crafters are attuned to how stitches are made and how they lie together to create a piece. Both mathematicians and crafters look for patterns, albeit in different forms. Yet students may not have experience attending to details of these kinds. Continuing in the vein of [2], because most students are novices, they are at the stage of only being able to notice or recognize the mathematical connections clearly pointed out to them by someone else, and are not yet at the stage of *marking*, where they can originate remarks about such observations. Recognizing that a shift in the level of noticing amongst the students will need to be a goal together with the mathematical and crafting goals is important as a teacher embarks upon the use of craft as a discovery-based tool in the classroom.

Although the additional layer of teaching involved in initiating students into the realm of coming-to-see mathematical ideas in the visual and kinesthetic world of craft may seem challenging, we argue that it significantly broadens students' utility with mathematics by giving them experience with transfer of knowledge from one domain to another.

To that end, here is a practical note. We recommend that students help each other, while the instructor surreptitiously listens in to make sure that their advice is correct. This builds confidence on the part of the student-as-instructor and reinforces the student's knowledge. In our experience, the student-as-receiver often finds the student-as-instructor easier to understand than the instructor, even if the student-as-instructor is using exactly the same words. Of course, if the advice is incorrect, this too yields valuable insight into the student's misunderstanding. This can allow for a gentle, yet powerful teaching moment.

5 How to Use This Book

While anyone can appreciate the visual aspects of this volume, it is more than a coffee-table book. (Still, if you like the look of any photo, the materials used and people involved are listed in the Credits section.) There are some special audiences for whom we have guidance on the best use of this book.

⋆ Crafters should note that the Crafting sections are written as ordinary patterns, so their formatting and language will be familiar. Every project was tested by at least one non-mathematician (who was not given access to the rest of the chapter), so we are confident that you can complete any project you attempt.

⋆ Students and parents seeking ideas for science-fair or research projects are directed to the Project Ideas subsections of most of the Teaching Ideas sections. There are also some ideas for independent work embedded in other parts of the Teaching Ideas sections.

⋆ Faculty who teach mathematics for liberal arts courses will find that fiber arts connections to mathematics provide powerful and salient examples for students to explore. (However, not all students observe these connections readily; see Section 4.) Open-ended investigations as suggested in the Teaching Ideas and Project Ideas sections can allow even beginning students to formulate research questions—a feat that they find impressive.

⋆ The Mathematics sections have technical levels that range from that of college mathematics majors to professional mathematicians. Readers hoping to enhance their mathematical expertise will find that persistence is their best approach, as is true for all mathematical learning.

⋆ We believe the careful treatment of topics in this volume (and [1]) will be of particular use to beginning graduate students in mathematics. However, professional mathematicians may also find this book to be a useful source for quickly acquiring background and details on subfields or concepts that are new to them.

In the previous volume we highlight nonorientability through homology [1, Chapter 1], frieze pattern nomenclature [1, Chapter 5], and surface curvature [1, Chapter 10]. Here we note wallpaper pattern identification and notation (Chapter 5), *m*-color planar patterns (Chapter 6), group actions (Chapter 6), and discrete spherical symmetry types (Chapter 8).

⋆ Of course, the readership closest to our hearts is mathematicians who craft. We have no advice for you in how to approach this material; you already know.

We also have a few observations to share about particular chapters.

Chapter 1 combines knitting with calculus to create circles and surfaces of revolution. Some instructional ideas have been tested in Calculus II classes: sarahmarie's students read an abbreviated version of the Mathematics section and she gave them a worksheet of the Designing a Cone questions. They had no trouble working through the problems until they encountered a question that specified a knitting gauge and head circumference and asked them to determine parameters a, b for the cone so that a wearable hat would result.

In Chapter 4, the educational activities are truly independent of educational level; the same activity is doable by sixth graders and graduate students alike (with different mental rewards). In the spirit of using many different crafts to create the same object, we note that the Chaos Game could be played using fabric paint dots instead of beads. Carolyn challenges readers to create a mola version of Sierpiński's triangle, and also to tat Pascal's triangle in multiple colors (avoiding cellular automata in the construction method, as Chapter 4 reveals is not useful).

Chapter 6 has two threads, one elementary and one advanced, throughout the Mathematics section. While the language is frequently technical, the reader can skip those parts, forge ahead, and see the concepts via example.

The Crafting section of Chapter 9 is characterized by unusual quilting techniques. Quilters are forewarned that several of the patterns are not for the algebraically faint of heart.

6   Conclusion

Gentle reader, we and the authors have endeavored to maintain a level of accessibility that will allow any reader to enjoy portions of the book. We do encourage you to stretch yourself by trying a new craft and learning some mathematics that is new to you (or new to the world). As you read and experiment more, you will appreciate the connections between the doing and learning of mathematics as and through craft.

Crafting by Concepts has been edited, indeed curated, to address many different audiences. We hope you find some part of this book that seems like it was written just for you.

sarah-marie belcastro and Carolyn Yackel, June 2010

Bibliography

[1] belcastro, sarah-marie and Yackel, Carolyn, eds. *Making Mathematics with Needlework*, Wellesley, MA, A K Peters, Ltd., 2008.

[2] Mason, J. "Researching From the Inside in Mathematics Education—Locating an I-You Relationship." In *Proceedings of the Eighteenth International Conference for the Psychology of Mathematics Education*, Vol. 1, edited by J. P. da Ponte and J. F. Matos, pp. 176–194. PME, Portugal, 1994.

[3] Conversations observed on *Ravelry*. Available at http://www.ravelry.com, September 2008–March 2010.

CHAPTER 1

knit knit revolution

AMY F. SZCZEPAŃSKI

1 Overview

For many knitters, the idea of knitting in the round brings to mind projects like socks, sweaters, hats, mittens, and gloves—anything that can be realized as modified cylindrical tubes. In addition to tubular projects, knitting in the round can also be used to knit shawls and other flat shapes. This chapter explores some of the mathematical techniques that can be used to design knitting patterns for flat circles and for large classes of shapes with circular cross sections. The mathematical and design techniques are not specific to knitting—all of these ideas work for crocheting shapes with circular cross sections as well. (A crocheter could simply replace every occurrence of the word "knit" with the word "crochet.") By relying on the simple relationship between the radius of a circle and its circumference, $C = 2\pi r$, patterns can be designed for many such shapes by following a fairly straightforward algorithm. Patterns are included for three hats whose designs showcase some of the mathematical techniques.

1.1 Circles that Lie Flat

One of the more famous applications of this mathematical relationship is Elizabeth Zimmermann's Pi Shawl [6]. In this pattern, Zimmerman periodically directs the knitter to double the number of stitches. The distance between increase rounds also continues to double. In the pattern in [6], the knitter is instructed to cast on nine stitches, knit one round even, double the number of stitches in rounds 2, 6, 13, 26, 51, 100, ... The number of rounds worked even between increase rounds will continue to double. Following the pattern, there will be 576 stitches in round 100. This process can be expressed in the language of algebra, as described in Section 2.1.

Zimmerman herself recognizes that there is a mathematical aspect of this pattern. In [6] she writes:

Have you begun to see the well-known geometric theory behind what you have been doing? ... It's Pi; the geometry of the circle hinging on the mysterious relationship of the circumference of a circle to its radius. ... [I]n knitters' terms, the distance between the increase-rounds, in which you double the number of stitches, goes 3, 6, 12, 24, 48, 96 rounds, and so on to 192, 394, 788, 1576 rounds for all I know. Theory is theory, and I have no intention of putting it into practice, as I do not plan to make a lace carpet for a football field.

As Zimmerman notes, this pattern has a reasonable mathematical grounding. Her algebra is a little bit off; she does not count the increase round anywhere in her calculations. It is ironic that she observes that π is at work here, yet her pattern makes no use of the actual value of π.

An Internet search will reveal many knitters' takes on this pattern. It is also cited in [5], where it is described as the *rounds* method of knitting a circle. The plan for knitting a Pi Shawl is fairly simple to describe and does not require any fancy calculations on the part of the knitter. Additionally, if one is knitting in wool with a fairly open stitch pattern—as is common with lace—the finished shawl can be blocked to lie relatively flat.

If, however, one is working a fairly tight stitch pattern, especially over a great number of rounds, the Pi Shawl will start to show its limitations. As the knitter continues to work, there will be longer and longer stretches that are worked without increasing. These sections will yearn to be cylindrical, their stitches smushed close near the center and pulled taut near the edge as the knitter tries to coax the circle to lie flat. Doubling the number of stitches results in a significant increase in the amount of fabric, and it will tend to ripple and curl in the early rounds after the increase.

The reason that Zimmerman's pattern works as well as it does is because it recognizes a fundamental fact about the relationship between the radius of a circle and its circumference: if the radius is multiplied by a specific constant (in this case, 2), then the circumference must be multiplied by the same constant. (While Zimmerman's pattern doesn't follow this rule exactly, it's pretty close.) Mathematically, this is the idea of "doing the same thing to both sides of the equation." Her pattern neglects an important issue, though: the radius increases every round, so the circumference should also increase with every round.

In [5] this is addressed, to some extent, by offering the alternative *rays* method of knitting a circle. The pattern for the rays method instructs the knitter (working in garter stitch) to increase by 16 stitches every four rows. As suggested in the name, in the rays method, the increases are aligned in rays that extend from the center of the circle. If all of the increases are in the same place every round, this shape is more polygonal than circular. Depending on the number of rays, the location of the increases, and the aggressiveness of the blocking, the end result can be fairly round. One method of determining where to place the increases is to divide the number of stitches in the current round by the number of increases and then round down. This will tell you the number of stitches, s, to knit between evenly spaced increases. Choose a random number between 0 and s, and knit that many stiches before you do the first increase.

Both the rounds method pioneered by [6] and the rays method described by [5] can be improved by working more closely with the equation $C = 2\pi r$. When knitting in the round to make a circle that lies flat, the system of increase depends on the gauge. Measure the number of rows in 4" (10 cm) and the number of stitches in 4" (10 cm). Calculate $2\pi \cdot \frac{\text{number of stitches}}{\text{number of rows}}$. This is how many stitches we should increase every round. (This formula is explained in Section 2.1.) This won't be a whole number, so we'll need to round it to a whole number of stitches. If we always round up, we'll eventually have too many stitches; if we always round down, we'll eventually have too few stitches. This won't matter when making a modest-size project out of a reasonably stretchy yarn. If, however, the goal is to make a gigantic circle out of an unyielding medium: be warned. We call this method of adding $2\pi \cdot \frac{\text{number of stitches}}{\text{number of rows}}$ stitches every round the *modified rays method*.

Both for the flat circle and for all of the other shapes described in this chapter there is a question of where to place the increases. Everywhere an increase is placed, the knitting will want to form a corner. For example, if there are four increase stitches each round, and they are evenly spaced about the round (thus lining the increases in each round up with the increases from the previous round), the shape will try very hard to have four corners. To make shapes that are as smooth as possible—especially when working a solid-colored pattern—increases should be scattered throughout the round, so that they do not line up with the increases from the round just below. When working a more complicated stitch pattern or color pattern, there may be fewer options for where to place the increases.

1.2 Surfaces of Revolution

Nothing stops us from taking this simple idea and extending it to a broader class of surfaces that are described by circles. In particular, surfaces of revolution have circular cross-sections. The obvious shapes to start with are spheres and tori (doughnut shapes), but nothing stops us from designing the patterns for any shape in this broader category—anything that is made up of circular slices. These sorts of shapes may interest the knitter making Christmas ornaments (especially if one wants to make knitted covers for the styrofoam balls from the craft store) or knitting models of food (such as doughnuts or ice cream cones). Additionally, these techniques can be applied to hats.

The only limit on the style of a knitted hat is the imagination of the knitter. Knitting a cylindrical tube (standard knitting in the round) and topping it with a flat circle makes a pillbox hat. Watch caps and other standard hats are cylinders topped with hemispheres. (Take a good look at someone's head; it's roughly shaped more like a silo than like a ball.) The stocking cap (or dunce cap) is a cone. In Section 4 there are three patterns for hats. One is a stocking cap knit in the standard way. The other two are hats with hemispheric caps knit back and forth on straight needles.

Mathematically, the problem of knitting any shape with circular cross-sections is exactly the same as knitting the flat circle: for each circular round of the shape, figure out the radius of the circle, calculate its circumference, and knit enough stitches in the round so that the circle is the right size. This is explained in mathematical detail in Section 2.2.

When knitting these shapes, many of the same questions come up as did with the flat circle. Again there is the issue of where to place increases and decreases. As with the flat circle, if increases (or decreases) line up, a ridge will be formed. Unless one is knitting a very large shape (or using very small stitches), it's best to try to scatter the increases and decreases about the round.

2 Mathematics

The construction of both flat circles and surfaces of revolution relies on a single geometric formula, that of the circumference of a circle. This is the familiar $C = 2\pi r$. Throughout this section h will stand for the height of a knit stitch, and w will stand for the width of a knit stitch.

2.1 The Flat Circle

The first application of $C = 2\pi r$ to knitting is the flat circle. The idea behind knitting a flat circle is to cast on a few stitches at the center of the circle and then to spiral out, increasing the number of stitches in each round. When knitting something circular, adding too few stitches per round will result in positive curvature (like a hat), and adding too many stitches per round will

result in negative curvature (like kale). The goal is to determine how many stitches should be added each round to achieve the proper growth rate. Therefore, when knitting in the round to make a circle that lies flat, we must calculate the number of stitches to increase each round so that the following round has the proper circumference.

The radius of round n will be nh (n times the height of a knit stitch), so the circumference of round n will be $2\pi nh$. To determine how many stitches to add when going from round n to round $n + 1$, we compare the circumferences. The desired circumference of round $n + 1$ is given by $2\pi(n + 1)h = 2\pi nh + 2\pi h$. Therefore, the difference between the circumferences of round $n + 1$ and round n is $2\pi h$. (This is essentially finite difference calculus: note that the derivative $\frac{dC}{dn} = 2\pi h$.) Dividing this by the width of a knit stitch, this means that for each round, one should increase by $\frac{2\pi h}{w}$ stitches. This is the modified rays method described on page 8.

The number $\frac{2\pi h}{w}$ will rarely be an integer, so the question of rounding arises. If $\frac{2\pi h}{w}$ is very close to an integer, then round and everything will work out. However, if its decimal part is near $\frac{1}{2}$, it's best to sometimes round up and sometimes round down—especially for a large project.

In contrast, the rays method is a simplification of the modified rays method. The rays method will always add 16 stitches every four rounds—an average of four stitches per round. If there are c stitches cast on, then in round r there will be $c + 4(r - 1)$ stitches. This method is independent of gauge. If the gauge is such that $\frac{2\pi h}{w}$ is really close to 4, then the rays method will produce results nearly identical to those of the modified rays method. Each method has the knitter add stitches at a constant rate per round. In the improved (modified rays) method, if there are c stitches cast on, then in round r there will be $c + (r - 1)\frac{2\pi h}{w}$ stitches.

With both of these methods, obtaining a result that is more round than polygonal relies on the placement of the increases. Whenever possible, try to place the increases randomly throughout the round.

Finally, let's look at Elizabeth Zimmermann's formula. As given in [6], her Pi Shawl pattern does not follow the algebraic equation to which she refers, $C = 2\pi r$. If we let r_i be the number of the first increase row, the correct placement of the increase rounds that double the stitch count *should* be at rows $r_i, 2r_i, 2^2 r_i, \ldots, 2^m r_i$. The reason for this is the old maxim from algebra class: whatever is done to one side of the equation must be done to the other side of the equation. To double the circumference, one must also double the radius. The circumference is defined by the number of stitches, and the radius is defined by the number of rounds. Therefore, each increase round should be numbered as twice the previous increase round.

This is not what the actual pattern directs, although it is close. Zimmerman's pattern has increases in rows 2, 6, 13, 26, 51, 100, ... The difference between Zimmerman's pattern and the correct algebra arises because she neglects the fact that the increase round itself adds to the radius of the circle. It appears that additionally, Zimmerman doubled the number of rows between increase rounds instead of doubling the round numbers in which increases occur:

$$6 = 2 + 3 + 1 = 2 + 3 + 1,$$
$$13 = 6 + 6 + 1 = 6 + 2 \cdot 3 + 1,$$
$$26 = 13 + 12 + 1 = 13 + 2^2 \cdot 3 + 1,$$
$$51 = 26 + 24 + 1 = 26 + 2^3 \cdot 3 + 1,$$
$$100 = 51 + 48 + 1 = 51 + 2^4 \cdot 3 + 1.$$

You will notice that twice the number of each increase round is usually close to, but not exactly equal to, the number of the next increase round:

$$6 = 2 \cdot 2 + 2,$$
$$13 = 2 \cdot 6 + 1,$$
$$26 = 2 \cdot 13 + 0,$$

$$51 = 2 \cdot 26 - 1,$$
$$100 = 2 \cdot 51 - 2.$$

The slight differences accumulate. With each increase round, the relationship between Zimmerman's radius and her circumference gets a little more out of whack. Of course, after working 100 rounds and having 576 stitches, the error is small enough so as not to be noticeable.

The location of the increase rounds in Zimmerman's original pattern is governed by the difference equation $A_{n+1} = 2 \cdot A_n + 3 - n$, where A_i is the round number of an increase round, and A_1 is defined by the pattern to be round 2. We solve the difference equation by attempting a solution of the form $A_n = a \cdot 2^n + bn + c$. We write A_{n+1} two ways, as $a \cdot 2^{n+1} + b(n+1) + c = 2(a \cdot 2^n + bn + c) + 3 - n$, and solve for b and c. We then use $A_1 = 2$ to solve for a. The closed form of this difference equation is $A_n = 3 \cdot 2^{n-1} + n - 2$.

We will now fix Zimmerman's formula to give a mathematically accurate description of the circle and call this the *Modified Pi Shawl*. To do this, we will refocus on rows when increases should happen. The stitch count should double only when the number of rounds has doubled. Therefore, if the first increase round is round 2, then there should also be increases in rounds 4, 8, 16, 32, 64, etc. Similarly, if the first increase is in round 3, then there should also be increases in rounds 6, 12, 24, 48, 96, etc. Instead of doubling the space between the increase rounds, one should simply double the round number. As before, in each increase round the number of stitches should double.

In this case, the location of the increase rounds and the number of stitches in a round are fairly straightforward to calculate. If we begin the doubling process in round r_i, we will double the stitch count in rounds $r_i, 2r_i, 2^2 r_i, 2^3 r_i, \ldots$. If we cast on c stitches, in each of the doubling rounds we will end up with $2c, 2^2 c, 2^3 c, 2^4 c, \ldots$ stitches. From here it is fairly straightforward to calculate how many stitches are in any given round r. The strategy is to determine how many times there has been an increase round and then multiply c by the suitable power of two.

As the round number of the increase rounds continues to double in this modified version of the Pi Shawl, finding their location will be done by taking logarithms base 2 to determine how many doublings there have been. To calculate the number of stitches in round r, we start by determining how many doublings there have been. The quantity $\frac{r}{r_i}$ will be exactly equal to a power of two in each of the increase rounds, by construction. Calculating $\log_2\left(\frac{r}{r_i}\right)$ will be a whole number in the doubling rounds and a decimal in other rounds; it counts the number of times the stitch count has been doubled before starting the row. Thus, we need $\left\lfloor \log_2\left(\frac{r}{r_i}\right) \right\rfloor + 1$ to tell us the number of times we have doubled by the end of round r. (Recall that $\lfloor x \rfloor$ is the greatest integer that does not exceed x.) So if we cast on c stitches and begin the doubling process in row r_i, then the number of stitches in row r (where $r \geq r_i$) will be $c \cdot 2^{\left\lfloor \log_2\left(\frac{r}{r_i}\right) \right\rfloor + 1}$.

We have described four methods for knitting a circle that lies flat. The modified rays method is the only one that relies on the number π and that takes into account the gauge of the knitting. It directs the knitter to add roughly $\frac{2\pi h}{w}$ per round, distributed as randomly as possible around the round. The other three methods do not involve the number π, nor do they take into account the gauge. They are the rays method from [5] that adds 16 stitches every four rounds (or four stitches each round), the Pi Shawl as described in [6], and the Modified Pi Shawl where the round number of the increases doubles rather than the space between the increases. We note, however, that for a lot of common yarns the number of stitches to add each round using the modified rays method with stockinette stitch comes out pretty close to the four prescribed by the rays method.

We prefer the Modified Pi Shawl to the Pi Shawl because it is more accurate from an algebraic point of view, and the arithmetic is nicer. Hence, we compare the remaining two of the three gauge-independent methods here (Modified Pi vs. rays). The number of stitches in the project will depend on which method one chooses as well as the number of stiches cast on. If one starts by casting on nine stitches, as is done in the original Pi Shawl, the number of stitches in each round of the Modified Pi Shawl far outnumbers the number of stitches in the rays method, as is shown in Table 1. We assume that we are adding four stitches each round in the rays method. From the beginning, the Modified Pi Shawl has greater than or equal to the number of stitches in each round as the rays method, and the difference continues to grow at a rate of $6(2^n) - 5$ (as derived from the On-Line Encyclopedia of Integer Sequences [4]).

round	Modified Pi Shawl	rays
1	9	9
3	18	17
6	36	29
12	72	53
24	144	101
48	288	197

Table 1. Number of stitches in selected rows when casting on nine stitches.

However, if one starts out by casting on a smaller number of stitches, then the rays method may outpace the Modified Pi Shawl. There is an interesting pattern here. As seen in Table 2, the differences between the numbers of stitches in the first five increase rounds in the rays method and in the Modified Pi Shawl are 3, 5, 9, 17, and 33, or more generally $2^n + 1$ (sequence A000051 in the OEIS [4]). Differences between adjacent terms in this sequence are consecutive powers of two.

This situation brings to mind the three bears from the fairy tale. With one initial number of stitches the Modified Pi Shawl dominates the rays. With another

round	Modified Pi Shawl	rays
1	5	5
3	10	13
6	20	25
12	40	49
24	80	97
48	160	193

Table 2. Number of stitches in selected rows when casting on five stitches.

initial number of stitches, the rays dominate the Modified Pi Shawl. Is there a number of stitches that is "just right"? The answer turns out to be yes, by casting on six stitches. Table 3 suggests that for the increase rounds in the Modified Pi Shawl (rounds 3, 6, 12, 24, ...), the number of stitches in the Modified Pi Shawl is two less than the number of stitches in the rays method.

round	Modified Pi Shawl	rays
1	6	6
3	12	14
6	24	26
12	48	50
24	96	98
48	192	194

Table 3. Number of stitches in selected rows when casting on six stitches.

Could it be a coincidence that the numbers work out this close for small-valued rounds? We can check this by looking at the expressions for the number of stitches in each round. Casting on six stitches and adding four stitches each round, we will have $6 + 4(r - 1)$ stitches in round r. In the Modified Pi Shawl, we will have $6 \cdot 2^{\lfloor \log_2(\frac{r}{3}) \rfloor + 1}$ stitches. If we restrict ourselves only to the increase rounds, we can omit the $\lfloor$ and $\rfloor$. We wish to see whether these two expressions always differ by two. That is, we need to verify the equation $6 \cdot 2^{\log_2(\frac{r}{3}) + 1} = 6 + 4(r - 1) - 2$. Sure enough, $4r = 6 \cdot \frac{r}{3} \cdot 2 = 6 \cdot 2^{\log_2(\frac{r}{3}) + 1}$ and $4r = 6 + 4(r - 1) - 2$. Therefore, in the case where we cast on six stitches, there is a balance between the

two methods. Let c be the number of stitches cast on in either method, let r_i be the location of the first doubling round in the Modified Pi Shawl, and let m be the number of stitches added each round in a generalized rays method. Then there will always be this balance between the number of stitches when $c = \frac{r_i}{2}m$. If r is an increase round in the Modified Pi Shawl and $c = \frac{r_i}{2}m$, then a round in the Modified Pi Shawl will have exactly $m - c$ more stitches than the same-numbered round in the rays method.

2.2 Surfaces of Revolution

The fundamental equation $C = 2\pi r$ is also used when designing patterns for surfaces of revolution. These shapes frequently come up in calculus classes; often the student is given an equation describing a graph in the plane and is asked to rotate this about an axis of revolution, generating a solid whose volume or surface area must be calculated.

The general method for creating a knitting pattern for such shapes is to imagine the original graph in the plane as made up of line segments each the length of one knit stitch. (This is somewhat like the approximation made when deriving the formula for arc length.) See Figure 1. Each of these knit stitches laid along the graph will be "rotated" about the axis of revolution by knitting a row with $\frac{2\pi x_i}{w}$ stitches, where x_i is the distance from the top of the stitch to the axis of rotation.

The design of revolutionary patterns can be reduced to the question of how best to fit these polygonal segments to the original curve. For one, there is no guarantee that the height of a knit stitch will fit a whole number of times along the curve being described. Obviously, approximations to the curve by smaller-height knit stitches will be more accurate.

Additionally, it is not always easy to figure out where the endpoint of each segment will be. Given a point (x_i, y_i) on the curve, the point (x_{i+1}, y_{i+1}) will also be on

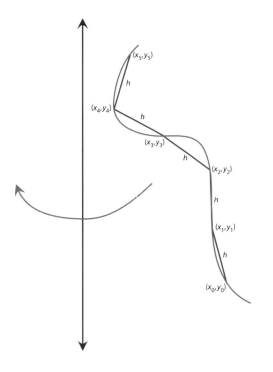

Figure 1. Let h be the height of a knit stitch and w be the width of a knit stitch. To knit a surface of revolution, identify points on the curve separated by a distance of h, cast on a round $\frac{2\pi x_0}{w}$ stitches, and then knit rounds of $\frac{2\pi x_i}{w}$ stitches for $i > 0$.

the curve, a distance h away from (x_i, y_i). Mathematically, this is equivalent to finding the points of intersection between the original curve and $(x - x_i)^2 + (y - y_i)^2 = h^2$, the circle of radius h centered at (x_i, y_i). For some shapes, it is easy to find an explicit formula for the next vertex in this polygonalization. For others, the most efficient strategy is to use technology to solve the equation. When the curve being rotated is piece-wise linear and/or quadratic, there is a shot at working with the equation by hand. However, if the curve is any more complicated than that, one is probably best off using technology—except for the few examples that invite the connoisseur to show off mastery of sixteenth century algebra.

2.3 An Algorithm for Designing Patterns

To create a pattern for a surface of revolution, our algorithm takes the curve to be rotated and approximates it by a polygonal curve the length of whose segments is equal to the height of one knit stitch; it then calculates $2\pi r$ for each vertex of the polygonal curve and divides this by the width of a knit stitch to find out how many stitches should be in the corresponding round.

The detailed algorithm is as follows:

1. Determine the gauge. Find the height h of a knit stitch and the width w of a knit stitch.

2. Identify the equation of the curve to be rotated. Denote this curve by γ. Find the location of γ with respect to the axis about which it will be rotated. For simplicity, these directions will assume that γ is rotated about the line $x = 0$.

3. On γ, select a point to be the starting point and denote it by (x_0, y_0).

4. On γ, select an ending point, (x_f, y_f).

5. Cast on $\frac{2\pi x_0}{w}$ stitches and join to knit in the round. When γ is a closed curve, a provisional cast-on is handy.

6. Knit the next round:

   (a) Let the current point of γ be (x_i, y_i).

   (b) Find the point (x_{i+1}, y_{i+1}) as a simultaneous solution to the equation $(x - x_i)^2 + (y - y_i)^2 = h^2$ and the equation for γ. As this system will have multiple solutions, choose the point that is not (x_{i-1}, y_{i-1}). If γ has self-intersections or other unusual features, one may need to develop more careful checks to ensure (x_{i+1}, y_{i+1}) is chosen correctly. Alternatively, this point can be found parametrically: Increase the value of the parameter t until the Euclidean distance from the point (x_i, y_i) to the point $(x(t), y(t))$ is h. At this value of t, $(x(t), y(t))$ gives the values of (x_{i+1}, y_{i+1}).

(c) Knit the next round with $\frac{2\pi x_{i+1}}{w}$ stitches.

(d) Repeat from (a) until the distance from (x_{i+1}, y_{i+1}) to (x_f, y_f) is less than or equal to $\frac{h}{2}$.

7. Either cast off or graft the beginning to the end.

Notice that this algorithm does not specify what to do with fractional stitches. Depending on γ and the gauge of the knitting, this may or may not be an issue.

2.4 Rotating Nice Shapes

Nicer calculations can be worked out when γ is a line or circle. Line segments are easy to approximate by shorter segments of length h. Furthermore, a circle is fairly easy to approximate as a regular n-gon with side length h. The ease of describing these types of shapes makes it simple to generate patterns for cones, spheres, and tori. The process for designing a cone is outlined in Section 3.3. Here we focus on spheres formed by rotating the semicircle $y = \pm\sqrt{r^2 - x^2}$ about $y = 0$ for $0 \leq x \leq r$, and tori formed by rotating the circle $x^2 + y^2 = r^2$ about an external axis.

When knitting an entire sphere, it's easiest to start by making a hemisphere; work from the equator to the pole, pick up stitches along the equator, and work to the other pole. When knitting a torus, we can start at any point on the circle, knit until we reach the point where the end would meet the beginning, and then graft the end to the beginning. As such, the descriptions of the torus and the sphere will both assume that the knitter starts by considering the point $(r, 0)$.

We now approximate the circle by a regular n-gon starting with the point $(r, 0)$, using rotation by $\frac{2\pi}{n}$ to locate the vertices. This rotation can be realized either with a rotation matrix or by describing the circle with equations parametrized by the central angle.

An n-gon inscribed in a circle can be divided into n identical isosceles triangles, with the apex of each triangle at the center of the circle. See Figure 2. Two sides of the triangle will be radii of the circle; the third will be one of the sides of the n-gon. When making a knitted object, the side-length of the n-gon (h, the height of a knit stitch) and the radius of the circle are specified. We need to calculate n, the number of sides, as this determines the number of rounds to be knit and how to change the number of stitches from one round to the next.

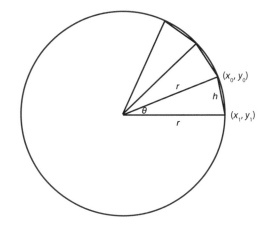

Figure 2. The arc from (x_0, y_0) to (x_1, y_1) can be approximated by a stitch of length h. To approximate the circle by a regular n-gon whose sides are of length h, set up an isosceles triangle with sides of length r, r, and h.

Additionally, assume that the height of a knit stitch is substantially shorter than the radius of the circle. For example, if the height of a knit stitch is $\frac{1}{4}''$, then it would be extremely difficult to knit a recognizable sphere with a $\frac{1}{4}''$ radius. This assumption allows us to say that $0 \leq \theta \leq \frac{\pi}{2}$.

An easy way to calculate n uses the law of cosines. If h is the height of a knit stitch, r is the radius of the circle, and θ is the central angle, then θ must satisfy the equation $\cos\theta = \frac{2r^2 - h^2}{2r^2}$.

Not every stitch height can be the side-length of a regular *n*-gon that is inscribed in a circle with a specified radius. When working with a fairly large gauge—or a fairly small circle—the results are better if the size of the circle is adjusted so that it will require an integral number of rounds. There are several ways to do this. One is to determine $\frac{2\pi}{\theta}$ and adjust θ so that $\frac{2\pi}{\theta} \in \mathbb{Z}$. From there the number of rounds can be recalculated. Another method is to adjust the radius of the circle (before calculating θ) so that the circumference is a multiple of *h*. Because we're restricted by the gauge to an integral number of sides, the polygon is unlikely to be exactly inscribed in, or circumscribed about, the circle.

Once the radius of the circle has been finalized, there are two workable strategies for finding the *x*-values of points along the circumference of the circle (still centered at the origin). One way to find these *x*-values is to set up parametric equations of the form $x = r\cos t$ and $y = r\sin t$. Next find the *x*-values for $t = k\theta$ where $0 \le k\theta < \frac{\pi}{2}$ for the circle and $0 \le k\theta < 2\pi$ for the torus. Another way to locate the points along the circle is by the use of a rotation matrix of the form

$$\begin{pmatrix} \cos\theta & \sin\theta \\ -\sin\theta & \cos\theta \end{pmatrix}.$$

Begin with $(r, 0)$ and multiply it by the matrix to get the next point on the circle. Iterate until reaching the desired endpoint.

Notice that neither method described immediately above requires explicit knowledge of θ. Both methods rely only on $\sin\theta$ and $\cos\theta$ to solve the equation $\cos\theta = \frac{h^2 - 2r^2}{2r^2}$ for θ. Because $0 \le \theta \le \frac{\pi}{2}$, both $\sin\theta$ and $\cos\theta$ are uniquely determined by the law of cosines and the identity $\sin^2\theta + \cos^2\theta = 1$. Additionally, using $t = k\theta$ merely means computing $\cos k\theta$ and $\sin k\theta$. There is no need to solve explicitly for θ.

After the polygonal vertex points are located, it is straightforward to determine how far they are from the axis of rotation and to calculate the length of the row to

be knitted, as was described in the general algorithm in Section 2.3.

2.5   Implementation

The algorithm in Section 2.3 can be implemented according to the *Mathematica* code below [3]. Stitches takes four arguments; these are the height of a knit stitch, the width of a knit stitch, the *x*-value of the initial point, and the *x*-value of the final point. The variables indicate the initial and final values of *x* and *y* on the curve as xi, xf, yi, and yf. The current and previous values of *x* and *y* on the curve are named xc, xp, yc, and yp.

This code uses FindRoot to solve the equation—which, in turn, uses Newton's method. By seeding Newton's method with xc + h, it starts close to the root being sought and seems to always find the desired point. To be extra careful, one could introduce a loop that would continue to evaluate Newton's method with different seed values until the solution found was different from xp. The code outputs a list of the number of stitches in each round (by taking the floor) and a list of points found along the curve.

```
Stitches[h_, w_, xi_, xf_] :=
    Module[{yi, yf, xc, yc, xp, yp, ans, ans2, a},
        f[x_] := Sqrt[x]; (*Enter function here*)
        (*Set up the result lists*)
        ans = {2*Pi*xi/w};
        ans2 = {};
        (*Define y-initial and y-final*)
        yi = f[xi];
        yf = f[xf];
        (*Set up initial values for the later loop*)
        xc = xi;
        yc = yi;
        xp = xi;
        yp = yi;
        (*This returns a rule:  x -> a*)
        While[(xf - xc)^2 + (yf - yc)^2 >= h^2,
            a = FindRoot[(x - xc)^2 + (f[x] - yc)^2
                    == h^2, {x, xc + h}];
            (*Move x-current to x-previous*)
            xp = xc;
```

```
yp = yc;
(*Apply the rule x -> a to get a value
    and increment x-current*)
  xc = x /. a;
  yc = f[xc];
  ans = Append[ans, Floor[2 Pi*xc/w]];
  ans2 = Append[ans2, {xc, yc}];
  ];
{ans, ans2}
];
```

An alternative approach would be to consider the rotated curve in terms of parametric equations. (Any function can be viewed parametrically by letting $x = t$ and then defining y in terms of t instead of x.) In this case, the starting point and ending point would be specified by beginning and ending values of t. The algorithm would then step through tiny values of t until the point $(x(t), y(t))$ was a distance h away from the previous point.

The patterns given in Section 4 were designed so that odd rows would always be stockinette, as some knitters have an aesthetic bias against stacked increases/decreases or against increasing/decreasing on purl rows. To produce a knitting pattern with odd rows plain, simply shift the increases/decreases to the even rows. That is, make sure that all of the even (or odd) numbered rows/rounds have the right number of stitches; the finished object will be sufficiently accurate and visually pleasing.

3   Teaching Ideas

While knitted surfaces of revolution can be used as models to introduce elementary-school children to common three-dimensional shapes, there are several ways for students familiar with algebra, geometry, or calculus to apply these ideas. The height of a knit stitch will continue to be denoted by h and the width of a knit stitch by w; all rotations are assumed to be about $x = 0$.

The subsequent projects and questions can be recast in terms of mosaics with regular tiles for situations where no knitted models are available (or for students who claim to have no interest in knitting). Simply replace stitches with tiles, and be aware that a few other adjustments may be needed. Other interpretations are possible, too: a problem about designing a pattern for knitting a hemisphere can be easily converted to a problem about covering the dome of the capitol building with solar panels.

3.1   Geometric Formulas and Estimation

Students who are learning about geometry at any level can use surfaces of revolution to explore standard geometric formulas.

The circle. Here are some questions that students can answer about knitted circles. They progress from those appropriate for K–3 level to high-school level. Of course, a teacher may need to translate the language of the questions to be appropriate for her classroom. Make sure that there are enough circles available, so that at most three students must share each one.

⋆ What is the radius (or the diameter) of the circle in inches? In centimeters? Can you measure it by counting stitches?

⋆ What is the circumference of the circle? Can you calculate it with the formula $C = 2\pi r$? Can you measure it with a ruler? Can you measure it by counting stitches?

⋆ If you measure the radius by counting stitches and then calculate the circumference, you will get a different answer than if you found the circumference by counting stitches. Why does this happen?

⋆ How many stitches do you think were needed to make the circle?

⋆ What is the area of the circle? Can you find it by measuring the radius and then calculating $A = \pi r^2$? Can you find the area in square inches? Square centimeters?

* What is the area of one stitch in square inches? In square centimeters?

* Can you find the total number of stitches used? How much yarn does it take to make one stitch? How much yarn was needed to knit the entire circle?

Surfaces of revolution. The above questions about the circle can be adapted to surfaces of revolution. While this topic is usually first addressed in calculus, the basic concepts can be introduced without reference to calculus.

* How was this shape formed as a surface of revolution? What curve or shape was rotated? About which axis?

* For any fixed point on the model, one can ask: How far is this point from the axis of rotation? In inches? In centimeters? What is the circumference of the model at this point? How long is it in inches? Centimeters? Number of stitches?

* Do we know a formula for finding the surface area of this shape? (Depending on the course, the answer to this question will vary!) Can you calculate the surface area of the shape? Can you approximate the surface area based on the number of stitches it has?

Students who are studying calculus can examine these ideas in more detail. The students can be challenged to adapt the design process for these shapes to come up with a set of instructions for knitting a set of cylindrical shells that fill the solid of revolution. These can be used to approximate the volume with the standard "shells" method. Or, the instructor can do the massive amounts of knitting required to make a model that will help to introduce the shells method in class. This would be a good opportunity to take advantage of the natural curl of stockinette stitch—and the bulk of thick

yarns! Shells knit out of stockinette stitch will naturally curl into cylinders. A Wolfram Demonstration is planned to generate such knitting patterns; check to see if it has come out yet.

3.2 Calculus: Parametric Equations

Some shapes are easier to model in the form $y = f(x)$, while others are easier to use in parametric form $x = f(t)$ and $y = g(t)$. (And some shapes are difficult no matter what form is used!) Calculus students exploring these models of surfaces of revolution can consider the pros and cons of each way of describing the shapes.

The method given above in Section 2.4 for working with spheres and tori is based on rotating a circle about a line and by finding the endpoints of the stitches by working with the sine and cosine of the central angle. To redo this example with parametric equations, the students can describe the circle as $x = \cos t$ and $y = \sin t$. As before, the initial point would be $(x_0, 0)$. However, to find the next point parametrically, they would substitute x and y into the equation $(x - x_0)^2 + (y - 0)^2 = h^2$ and solve for t. For later points, they would plug x and y into $(x - x_i)^2 + (y - y_i)^2 = h^2$. This is more difficult algebraically than the method of using the law of cosines; however, these equations can be solved fairly routinely by using a computer algebra system.

This method can be extended to any curve γ. The students can view γ as $y = f(x)$ or as $x = f(t)$ and $y = g(t)$. Once they know the coordinates of a point (x_i, y_i), they can calculate the next point by solving (nonparametrically) $(x - x_i)^2 + (f(x) - y_i)^2 = h^2$ for x (and then plugging in to find y) or by solving (parametrically) $(f(t) - x_i)^2 + (g(t) - y_i)^2 = h^2$ for t (and then plugging in to find x and y). Alternatively, when the curve is described parametrically, they can solve the problem numerically by allowing the value of t to gradually increase until $(f(t), g(t))$ is a distance h away from (x_i, y_i).

Figure 3. College students work in groups on a worksheet adapted from Section 3.3.

The students can choose a pattern for some other shape (like a paraboloid) and construct the pattern both by following the algorithm in Section 2.3 and trying to come up with both a parametric and nonparametric description.

3.3 Designing the Cone

Students possessing facility with algebra should be able to work out the pattern for making a knitted cone, as shown in Figure 3.

To form the cone, rotate a line segment around an axis of revolution and add a knitted circle as base. For simplicity, define the line segment by its endpoints $(a, 0)$ and $(0, b)$ and assume that the rotation is about the y-axis (see Figure 4).

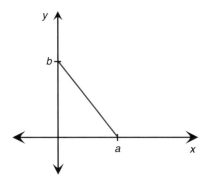

Figure 4. A cone can be realized by rotating the line segment connecting $(a, 0)$ and $(0, b)$ about the y-axis.

In order to produce a pattern with whole numbers, both a and $\sqrt{a^2 + b^2}$ should be multiples of h. One of the easiest ways to accomplish this is to choose a suitable Pythagorean triple and multiply each of the three numbers by h.

Generating the pattern comes down to the following steps:

1. The radius of the base of the cone is a. How many rounds must you knit to make a circle of radius a?

2. How many stitches must you increase each round when making a circle of radius a?

3. What is the slant height of the cone (the length of the segment from $(a, 0)$ to $(0, b)$)? How many stitches high is that? So, how many rounds will you knit when making the rest of the cone?

4. Find the equation of the line passing through $(a, 0)$ and $(0, b)$.

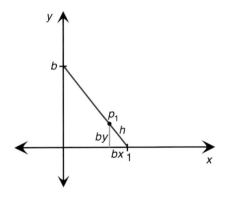

Figure 5. The height of a knit stitch, h, is the hypotenuse of a right triangle of side lengths Δx, Δy, and h. The quantities Δx and Δy are related by the slope of the line.

5. The Pythagorean theorem can be used to find a point on the line segment that is h units away from $(a, 0)$. See the diagram in Figure 5. The lengths of the sides of this triangle satisfy the equation $(\Delta x)^2 + (\Delta y)^2 = h^2$. Use what you know about the line segment to find $\frac{\Delta y}{\Delta x}$.

6. Solve for Δy and substitute this into the Pythagorean theorem.

7. Solve for Δx.

8. Calculate the x-coordinate of the point $p_1 = (a - \Delta x, \Delta y)$ in Figure 5.

9. How many stitches should be in the round containing p_1?

10. What do you do to find the number of stitches in the next round? In the one after that?

3.4 Working with Computer Algebra Systems

Students who are learning how to work with a Computer Algebra System (CAS) such as *Mathematica*, Maple, or Sage can write programs that generate patterns for surfaces of revolution. (See Section 2.5 for a *Mathematica* example.) The students can work from the algorithm given in Section 2.3. The algorithm in Section 2.3 can be viewed as pseudocode for the implementation, with step 6 as the pseudocode for the main loop.

4 Crafting Revolutionary Hats

4.1 Knitting up a Baby Bumble Bee Hat

The KuaBBBee Hat, shown in Figure 6, is knit from back to front by creating a hemisphere between two rectangles that are then grafted into a cylinder.

Materials

⋆ Cascade 220 yarn (100% Peruvian highland wool; 100g/3.5 oz., 220 yds.), 80 yds. ($\sim$ 1.2 oz.) of MC (black) color #8555 and 30 yds. ($\sim$ .5 oz.) CC (goldenrod) color #7827

or

Valley Yarns Northampton (100% wool; 100g/3.5 oz., 247 yds.) in black (07) and gold (15). If using one new ball of each color, there will be plenty of yarn left over.

Figure 6. A KuaBBBee hat.

* One 16" US #7 circular needle, or a longer #7 circular needle (or a set of #7 straight needles and a set of five #7 double-pointed needles).

* Two stitch markers.

* Scrap yarn for optional provisional cast-on.

* Yarn needle.

* Optional buttons or small circles of felt to use as eyes for the bumblebee and scrap yarn or heavy-duty thread to attach them.

Gauge

* 21 stitches and 28 rows in 4" by 4".

Size

* Child (1 yr.), 17" circumference.

Abbreviations

* K (or k) means knit.

* P (or p) means purl.

* M1 (or m1) means make 1. One way to do this is to lift the horizontal strand between two stitches onto the needle and knit the loop so it twists.

* Pm (or pm) means place marker.

* Sl m (or sl m) means slip marker.

* CO means cast on.

* K2tog (or k2tog) means knit two together.

* MC means main color.

* CC means contrasting color.

Instructions

When switching colors, carry the unused color along the side, trapping it loosely in the selvage. Resist the urge to slip the first stitch of the row; you will be picking up stitches along the edge later.

CO 24 sts with main color. (Optional: use provisional cast-on with scrap yarn, CO 24 sts, and knit one row with main color. If you do this, you will graft together the seam instead of sewing it, so leave a long tail to be used for grafting.)

Row 1 (and all odd-numbered rows): Purl every st, slipping markers as you encounter them.

Row 2: K10, pm, (k1, m1) 3 times, k1, place marker, k10. (7 sts between markers).

Row 4: K10, sl m, (k1, m1) 2 times, k1, (k1, m1) 3 times, k1, sl m, k10 (12 sts between markers).

Row 6: Change to CC. K10, sl m, k2, m1, (k3, m1) 3 times, k1, sl m, k10 (16 sts between markers).

Row 8: K10, sl m, k2, m1, (k4, m1) 3 times, k2, sl m, k10 (20 sts between markers).

Row 10: K10, sl m, k2, m1, (k5, m1) 3 times, k3, sl m, k10 (24 sts between markers).

Row 12: Change to MC. K10, sl m, (m1, k6) 4 times, sl m, k10 (28 sts between markers).

Row 14: K10, sl m, (k9, m1) 3 times, k1, sl m, k10 (31 sts between markers).

Row 16: K10, sl m, (m1, k10) 3 times, k1, sl m, k10 (34 sts between markers).

Row 18: Change to CC. K10, sl m, k6, (m1, k12) 2 times, m1, k4, sl m, k10 (37 sts between markers).

Row 20: K10, sl m, k18, m1, k19, m1, sl m, k10 (39 sts between markers).

Row 22: K10, sl m, k10, m1, k19, m1, k10, sl m, k10 (41 sts between markers).

Row 24: Change to MC. K10, sl m, k20, m1, k21, sl m, k10 (42 sts between markers).

Row 26: K10, sl m, k10, m1, k32, sl m, k 10 (43 sts between markers).

Row 28: K every st, sl m as you encounter them.

Row 30: Change to CC. K every st, sl m as you encounter them.

Row 32: K10, sl m, k20, k2tog, k21, sl m, k10 (42 sts between markers).

Row 34: K10, sl m, k2tog, k40, sl m, k10 (41 sts between markers).

Row 36: Change to MC. K10, sl m, k10, k2tog, k20, k2tog, k7, sl m, k10 (39 sts between markers).

Row 38: K10, sl m, k12, k2tog, k12, k2tog, k11, sl m, k10 (37 sts between markers).

Row 40: K10, sl m, k2tog, k15, k2tog, k16, k2tog, sl m, k10 (34 sts between markers).

Row 42: K10, sl m, k5, k2tog, k10, k2tog, k10, k2tog, k3, sl m, k10 (31 sts between markers).

Row 44: K10, sl m, k2tog, k13, k2tog, k12, k2tog, sl m, k10 (28 sts between markers).

Row 46: K10, sl m, k1, k2tog, (k6, k2tog) 3 times, k1, sl m, k10 (24 sts between markers).

Row 48: K10, sl m, (k4, k2tog) 4 times, sl m, k10 (20 sts between markers).

Row 50: K10, sl m, (k2tog, k3) 4 times, sl m, k10 (16 sts between markers).

Row 52: K10, sl m, k1, (k2tog, k2) 3 times, k2tog, k1, sl m, k10 (12 sts between markers).

Row 54: K10, sl m, k2tog 3 times, k2, k2tog 2 times, sl m, k10 (7 sts between markers).

Row 56: K10, sl m, k1, k2tog 3 times, sl m, k10 (4 sts between markers).

Row 57: Purl all stitches.

Graft together the seam at the front of the head (optional: bind off and sew the seam together) and sew together the seam at the back of the head (optional: graft it together if you used a provisional cast-on).

With MC and 16″ circular needles (or set of dpn) pick up and knit 80 stitches (that is, four stitches per six rows) around the base of the hat, and join for knitting in the round. Work in k2, p2 rib for 1.5" or desired length. Bind off. Weave in ends. If desired, attach eyes to front of hat.

4.2 Possibly Stashbusting Handspun Stripehead

This adult variant on the Baby Bee Hat, shown in Figure 7, was created by sarah-marie belcastro to use the small amounts of handspun yarn one acquires, or the bits of leftover fancy commercial yarn one can't bear to let go to waste.

Figure 7. A Stripehead hat.

Materials

★ Worsted handspun yarn, ∼ 1.25 oz. in each of two colors, or smaller amounts of more colors
or
Valley Yarns Amherst (100% Merino wool; 50 g, 109 yds.) or Colrain (50% Merino/50% Tencel; 50 g, 109 yds.).

★ One 24" US #8 circular needle, or a set of #8 straight needles and a set of five #8 double-pointed needles.

★ Two stitch markers.

★ Scrap yarn for optional provisional cast-on.

★ Yarn needle.

Gauge

★ 4 stitches per inch and 5.5 rows per inch.

Size

★ Adult, 22" circumference.

Abbreviations

★ K (or k) means knit.

★ P (or p) means purl.

★ M1 (or m1) means make 1. One way to do this is to lift the horizontal strand between two stitches onto the needle and knit the loop so it twists.

★ Pm (or pm) means place marker.

★ Sl m (or sl m) means slip marker.

★ CO means cast on.

★ K2tog (or k2tog) means knit two together.

* MC means main color.

* CC means contrasting color.

Instructions

When switching colors, carry the unused color along the side, trapping it loosely in the selvage.

CO 28 sts with MC: CO 12 sts, pm, CO 4 sts, pm, CO 12 sts. (Optional: use provisional cast-on with scrap yarn, CO 28 sts, and with main color k12, pm, k4, pm, k12. If you do this, you will graft together the seam instead of sewing it, so leave a long tail to be used for grafting.)

Row 1 (and all odd-numbered rows): K8, p4, purl stitches between markers, p4, k12.

Row 2: K12, pm, (k1, m1) 3 times, k1, pm, k12 (7 sts between markers).

Row 4: K12, sl m, (k1, m1) 2 times, k1, (k1, m1) 3 times, k1, sl m, k12 (12 sts between markers).

Row 6: Change to CC. K12, sl m, k2, m1, (k3, m1) 3 times, k1, sl m, k12 (16 sts between markers).

Row 8: K12, sl m, k2, m1, (k4, m1) 3 times, k2, sl m, k12 (20 sts between markers).

Row 10: K12, sl m, k2, m1, (k5, m1) 3 times, k3, sl m, k12 (24 sts between markers).

Row 12: Change to MC. K12, sl m, (m1, k6) 4 times, sl m, k12 (28 sts between markers).

Row 14: K12, sl m, (k9, m1) 3 times, k1, sl m, k12 (31 sts between markers).

Row 16: K12, sl m, (m1, K10) 3 times, k1, sl m, k12 (34 sts between markers).

Row 18: Change to CC. K12, sl m, k6, (m1, k12) 2 times, m1, k4, sl m, k12 (37 sts between markers).

Row 20: K12, sl m, k18, m1, k19, m1, sl m, k12 (39 sts between markers).

Row 22: K12, sl m, k10, m1, k19, m1, k10, sl m, k12 (41 sts between markers).

Row 24: Change to MC. K12, sl m, k20, m1, k21, sl m, k12 (42 sts between markers).

Row 26: K12, sl m, k10, m1, k32, sl m, k12 (43 sts between markers).

Row 28: Knit every stitch, slip markers as you encounter them.

Row 30: Change to CC. Knit every stitch, slip markers as you encounter them.

Row 32: K12, sl m, k20, k2tog, k21, sl m, k12 (42 sts between markers).

Row 34: K12, sl m, k2tog, k40, sl m, k12 (41 sts between markers).

Row 36: Change to MC. K12, sl m, k10, k2tog, k20, k2tog, k7, sl m, k12 (39 sts between markers).

Row 38: K12, sl m, k12, k2tog, k12, k2tog, k11, sl m, k12 (37 sts between markers).

Row 40: K12, sl m, k2tog, k15, k2tog, k16, k2tog, sl m, k12 (34 sts between markers).

Row 42: Change to CC. K12, sl m, k5, k2tog, k10, k2tog, k10, k2tog, k3, sl m, k12 (31 sts between markers).

Row 44: K12, sl m, k2tog, k13, k2tog, k12, k2tog, sl m, k12 (28 sts between markers).

Row 46: K12, sl m, k1, k2tog, (k6, k2tog) 3 times, k1, sl m, k12 (24 sts between markers).

Row 48: Change to MC. K12, sl m, (k4, k2tog) 4 times, sl m, k12 (20 sts between markers).

Row 50: K12, sl m, (k2tog, k3) 4 times, sl m, k12 (16 sts between markers).

Row 52: K12, sl m, k1, (k2tog, k2) 3 times, k2tog, k1, sl m, k12 (12 sts between markers).

Row 54: Change to CC. K12, sl m, k2tog 3 times, k2, k2tog 2 times, sl m, k12 (7 sts between markers).

Row 56: K12, sl m, k1, k2tog 3 times, sl m, k12 (4 sts between markers).

Row 57: K8, p12, k8.

Graft together the 24 stitches on the needle (optional: bind off and sew the seam together) and sew together the seam at the start of the work (optional: graft it together if you used a provisional cast-on). If you are feeling particularly clever, graft so that the garter border

Figure 8. Three Coney Hats on three friends.

appears uninterrupted. (This only requires doing regular Kitchener stitch.) Weave in remaining end of MC. Place hat on head.

4.3   Coney Hat

The carefully chosen colors (particular shades of orange, dark blue, and teal) used in this hat were inspired by Akiyoshi Kitaoka's optical illusions [1, 2]. The teal looks greenish when it is near the orange and it looks like a much brighter blue when it is next to the dark blue. The effect is exaggerated in bright sunlight and from a distance—when you look at the hat close up at night, it is not as illusory. Figure 8 shows three Coney Hats made of three different yarns, with varying degrees of optical illusion. The rightmost hat is made in the exact colors given in this pattern, the middle hat is made in a different yarn with close (but not exact) colors, and the leftmost hat is made in completely different colors and yarn.

Materials

★ Cascade Pastaza yarn (50% llama/50% wool; 100g/3.5 oz. 132 yds.). One skein each MC delft (also called starry night and midnight) #6002 (60g needed), CC1 teal #013 (35g needed), CC2 russet #010 (40g needed)

or

Valley Yarns Northampton yarn (100% wool; 100g/3.5 oz., 247 yds.). One skein each MC midnight heather #29 (60g needed), CC1 jade #34 (35g needed), CC2 sunset #16 (40g needed).

★ Set of five US #9 double-pointed needles.

Gauge

★ 18 stitches and 23 rows over 4″ by 4″.

Size

★ Adult medium, 22″ circumference.

Row	Color	# Dec	# Stitches
1	MC	0	88
2	CC1	0	88
3	MC	0	88
4	CC1	0	88
5	MC	0	88
6	CC1	0	88
7	MC	0	88
8	CC1	0	88
9	MC	0	88
10	CC2	0	88
11	MC	0	88
12	CC2	0	88
13	MC	0	88
14	CC2	0	88
15	MC	0	88
16	CC2	0	88
17	CC1	2	86
18	CC2	1	85
19	CC1	1	84
20	CC2	1	83
21	CC1	1	82
22	CC2	1	81
23	CC1	1	80
24	CC2	1	79
25	MC	1	78
26	CC2	1	77
27	MC	1	76
28	CC2	1	75
29	MC	1	74
30	CC2	1	73
31	MC	1	72
32	CC2	1	71
33	MC	1	70
34	CC1	1	69
35	MC	1	68
36	CC1	1	67
37	MC	2	65
38	CC1	1	64
39	MC	1	63
40	CC1	1	62
41	MC	1	61
42	CC2	1	60
43	MC	1	59
44	CC2	1	58
45	MC	1	57
46	CC2	1	56
47	MC	1	55
48	CC2	1	54

Row	Color	# Dec	# Stitches
49	CC1	1	53
50	CC2	1	52
51	CC1	1	51
52	CC2	1	50
53	CC1	1	49
54	CC2	1	48
55	CC1	1	47
56	CC2	1	46
57	MC	2	44
58	CC2	1	43
59	MC	1	42
60	CC2	1	41
61	MC	1	40
62	CC2	1	39
63	MC	1	38
64	CC2	1	37
65	MC	1	36
66	CC1	1	35
67	MC	1	34
68	CC1		33
69	MC	1	32
70	CC1	1	31
71	MC	1	30
72	CC1	1	29
73	MC	1	28
74	CC2	1	27
75	MC	1	26
76	CC2	1	25
77	MC	2	23
78	CC2	1	22
79	MC	1	21
80	CC2	1	20
81	CC1	1	19
82	CC2	1	18
83	CC1	1	17
84	CC2	1	16
85	CC1	1	15
86	CC2	1	14
87	CC1	1	13
88	CC2	1	12
89	MC	1	11
90	CC2	1	10
91	MC	1	9
92	CC2	1	8
93	MC	1	7
94	CC2	1	6
95	MC	1	5
96	CC2	1	4

Figure 9. This table is a graphic version of the textual hat instructions.

Abbreviations

* K (or k) means knit.

* P (or p) means purl.

* CO means cast on.

* MC means main color.

* CC1 and CC2 mean contrasting colors 1 and 2.

Notes

To avoid jogs in the stripes when knitting in the round, after knitting the last stitch of the round, slip the first stitch of the next round (purlwise). This moves the beginning of the round by one stitch. Be careful to leave plenty of slack in any floats carried on the wrong side of the work. Rounds will be numbered starting *after* the ribbing. See Figure 9 for a graphic version of the following textual instructions.

The color pattern is: MC alternating with CC1 (eight rounds), MC alternating with CC2 (eight rounds), CC1 alternating with CC2 (eight rounds), MC alternating with CC2 (eight rounds).

Rounds 1–15 odd, 25–47 odd, 57–79 odd, and 89–95 odd are worked in MC.
Rounds 2–8 even, 17–23 odd, 34–40 even, 49–55 odd, 66–72 even, and 81–87 odd are worked in CC1.
Rounds 10–32 even, 42–64 even, and 74–96 even are worked in CC2.

Instructions

CO 88 sts with the main color, and join for working in the round. Work for between $1\frac{1}{2}''$ and 2″ in k2 p2 ribbing. The next round will start counting with 1.

Rounds 1–16: Work even, knit every stitch. The first eight rows will be stripes alternating MC with CC1; the next eight rows will be stripes alternating MC with CC2.

Round 17: Knit every stitch, but over the course of the round decrease two stitches, placing the decreases arbitrarily (86 sts). The goal in placing the decreases arbitrarily is to avoid stacking the decreases from one row atop (or beneath) the decreases from nearby rows.

Rounds 18–36: Knit every stitch, but over the course of each round, decrease one stitch, placing the decreases arbitrarily (67 sts at the end of round 36).

Round 37: Knit every stitch, but over the course of the round decrease two stitches, placing the decreases arbitrarily (65 sts).

Rounds 38–56: Knit every stitch, but over the course of each round, decrease one stitch, placing the decreases arbitrarily (46 sts at the end of round 56).

Round 57: Knit every stitch, but over the course of the round decrease two stitches, placing the decreases arbitrarily (44 sts).

Rounds 58–76: Knit every stitch, but over the course of each round, decrease one stitch, placing the decreases arbitrarily (25 sts at the end of round 76).

Round 77: Knit every stitch, but over the course of the round decrease two stitches, placing the decreases arbitrarily (23 sts).

Rounds 78–96: Knit every stitch, but over the course of each round, decrease one stitch, placing the decreases arbitrarily (4 sts at the end of round 96).

Pull yarn through remaining four stitches. Weave in loose ends. Optional: make a pom-pom or tassel and affix to top of the hat.

Bibliography

[1] Kitaoka, Akiyoshi. "Akiyoshi's Illusion Pages." Available at http://www.psy.ritsumei.ac.jp/~akitaoka/index-e.html, 2010.

[2] Kitaoka, Akiyoshi. "Color Illusion." Available at http://www.psy.ritsumei.ac.jp/~akitaoka/color-e.html, 2005

[3] Conant, James. Personal communication, May 22, 2009.

[4] Sloane, Neil J.A. "The On-Line Encyclopedia of Integer Sequences." AT&T Labs Research. Avaliable at http://www.research.att.com/~njas/sequences/, 2010.

[5] Waterman, Martha. *Traditional Knitted Lace Shawls*. Interweave Press, Loveland, CO, 1998.

[6] Zimmermann, Elizabeth. *Elizabeth Zimmermann's Knitters Almanac: Projects for Each Month of the Year*. Dover Publications, Toronto, 1981.

CHAPTER 2

generalized helix striping

SARAH-MARIE BELCASTRO

1　Overview

Knitters have many ways to vary color in a project. The simplest (from the point of view of the knitter) involve using yarn that varies in color. One can acquire yarn with several shades of the same color, or with several colors, or even several shades of several colors. However, predicting how these colors will fall in a finished knitted project is quite difficult, and so a knitter who desires a particular color pattern usually uses separate balls of yarn in different colors.

Within the realm of colorwork in knitting, there are several techniques. *Intarsia* refers to blocks of color, where different colors of yarn are used in different regions of a given project. Only one ball of yarn is used at a time, and the knitter switches balls of yarn when switching solid-color regions (see Figure 1).

Stranded color knitting involves using multiple colors interspersed in a single region of a project, as shown in Figure 2. Here, multiple balls of yarn are used at once; the knitter switches colors frequently and the unused yarns are carried along behind the work. As we will see in this chapter, sometimes it is actually useful to knit with more than one strand of the same color of yarn, no matter how counterintuitive that seems. *Striping* is more an effect than a technique. Different knitting techniques are needed for horizontal, vertical, and diagonal stripes. In this chapter, we will consider a specific kind of diagonal stripe.

When multiple colors of yarn are used, no matter which technique is employed, the strands of yarn leading from the balls of yarn to the project-in-progress get tangled. Some of this is simply because the balls of yarn jounce around the knitter's bag (rare is the knitter who does not use a bag, for fear of escaped or pet/child-attacked balls of yarn), but there is also a mathematical process at work as the knitter switches from one ball of yarn to another. Twisted strands form a *braid*; mathematical braids were introduced in conjunction with knitted cables in [2], and the same mathematics is at work here. Carolyn Yackel mentioned several years ago that it would be interesting to study the braids formed by the strands tangling in colorwork. As a mathematical problem to study, that seems huge—there seem to be as many possible braids as colorwork patterns, and that's a lot. Here, we will restrict ourselves to a single

 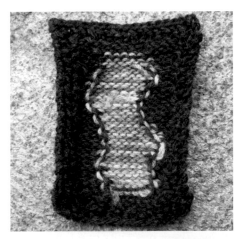

Figure 1. A sample of intarsia, with both the front of the work (left) and the back of the work (right) shown.

Figure 2. A sample of stranded knitting, with both the front of the work (left) and the back of the work (right) shown.

technique for producing stripes and see what insight we can gather. A practical reason to study the braids formed by yarn tangling is that understanding the braids can lead to efficient methods for untangling the yarn!

The motivation for this chapter's investigation was the publication of Joan Hamer's Helix Knitted Cap [4] in the 2007 Pattern-A-Day calendar. The idea of helix striping is much older—Hamer published the pattern online in 1997 and the technique shows up in Montse Stanley's *Knitter's Handbook* [5]. A *helix* is a particular type of curve that spirals through three dimensions. One could consider stitches from knitting in the round to travel along a helix, because while the last stitch of a given round is adjacent to the first stitch of the following round, it is also above the last stitch of the previous round. Ordinarily, stripes made when knitting in the round seem to "jog" at the switch of colors. The original helix striping construction is for jogless three-color striping when knitting in the round. Each color appears to form a continuous spiral around the work.

For a given number, k, of stitches in the round, one begins by casting on (making) one third, or $\frac{k}{3}$, of the stitches in each color. Then, one knits one third ($\frac{k}{3}$) of the stitches in the third color, drops the working yarn and picks up the free end of the first color, knits with

the first color until reaching the free end of the second color, drops the working yarn, picks up the free end of the second color to knit, and so forth. Aside from creating jogless stripes, this also has the advantage of being somewhat mindless; instead of thinking for a moment about which color to knit with next, or how many stitches to knit in a given color, this construction has a built-in cue for how many stitches in which color to knit when.

As we will see in Section 2, this process can be generalized easily to using any number of colors, and with a bit more creativity to any stripe height. (The height of a stripe is the number of rows at its greatest thickness.) Helix striping is always worked in the round. Knitting stripes with height greater than one involves using multiple strands of each color, which is irritating at the start of the construction because one has to make multiple balls of each color. But the knitting process is as easy as when knitting using a single strand of each color, and this makes it worth one's while. The reason the author decided to pursue Yackel's question about what sorts of braids are generated when doing colorwork knitting is that she was knitting using helix striping, and wondered whether there might be an efficient way to detangle the yarn strands! This question will be answered in Section 2.2.

In the process of rephrasing the question to be relevant to the specific context—pretty much the most important part of any mathematical research—it became apparent that there were not only questions about braids to pursue, but questions in number theory as well. These latter questions are about how many stitches should be knit in each color, how many balls of yarn should be used, and when it is possible to use a particular number of colors with a particular stripe height for a particular number of stitches in the intended project. It turns out that choosing a number of colors and a stripe height determines everything else; see Section 2.1 for details.

The modular arithmetic involved in Section 2.1 is particularly suitable for college-level number theory classes, but there are more basic questions that one can ask of upper-elementary or middle-school students. These are outlined in Section 3.1. Variants on the problem and some student project ideas are also discussed in Section 3.

Finally, Section 4.1 gives instructions for making a helix-striped nightcap in which the colors spiral to the point of the cap. No short rowing or knitting on the bias is involved. Section 4.2 explains how to knit a matching pair of bed socks. The bed socks could be reinterpreted as boot socks, with the nightcap reinterpreted as a fall or spring hat. The chapter closes with Section 4.3, which is a compendium of the knitting instructions for making jogless stripes of height h in n different colors.

2   Mathematics

In Section 1, the basic three-color, height-one helix striping construction is described on a round of k stitches. Here it is shown in Figure 3, where the view is from above and the knitting proceeds from the inside of the figure to the outside. (Knitting in the round literally produces a spiral, rather than a set of stacked short cylinders as is produced in crochet.) The meticulous reader will have noticed that after casting on $\frac{k}{3}$ stitches in color 3, one knits another $\frac{k}{3}$ stitches in color 3 before knitting $\frac{k}{3}$ stitches in color 1 and $\frac{2k}{3}$ stitches in color 2 ... that is, the stripe-segments are not of equal width.

That, sad to say, is aesthetically unpleasing (to the mathematician if not the knitter). We offer the following alternate construction, which achieves the same goals of joglessness and mindlessness as the original construction. A diagram of the construction is shown in Figure 4. Cast on $\frac{k}{2}$ stitches in colors 1 and 2; join to knit in the round. Knit in color 3 ($\frac{k}{2}$ stitches) until reaching the free strand of color 1; pick it up and knit ($\frac{k}{2}$ stitches) until reaching the free strand of color 2, and continue in color 2. Generically, knit until reaching the next free strand of yarn, pick up that strand, and repeat.

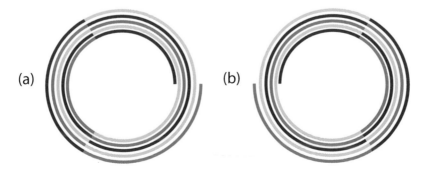

(a)   (b)

Figure 3. The original helix striping construction for three colors, in (a) left-handed and (b) right-handed versions.

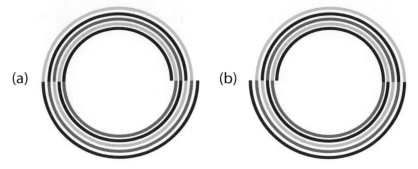

Figure 4. A revised helix striping construction for three colors, in (a) left-handed and (b) right-handed versions.

This is easily generalizable to n colors by casting on $\frac{k}{(n-1)}$ stitches in colors $1, 2, \ldots, n-1$; join to knit in the round. Knit in color n ($\frac{k}{(n-1)}$ stitches) until reaching the free strand of color 1; pick it up and knit ($\frac{k}{(n-1)}$ stitches) until reaching the free strand of color 2, and continue in color 2. Generically, knit until reaching the next free strand of yarn, pick up that strand, and repeat.

With a little bit more ingenuity, we can generalize to stripes of height other than one. The height-one stripes were formed by rows of stitches forming partial rounds of knitting, so that each stripe progressed at a constant (small) angle around the cylinder formed by the work. That is, the stripes approximated helices by piecewise intrinsically linear disjoint segments. Stripes of greater height still only approximate helices, but have a greater

angle of inclination to the horizontal and form continuous regions of color.

2.1 Generalized Helix Striping: The Number Theory

As a beginning example, suppose we want to use two colors (red and yellow) and obtain stripes of height two. Because the cast-on process is somewhat detailed, imagine we have cast on k stitches in scrap yarn. Our goal is to have the second layer of red stitches overlap the first layer of red stitches by half, and likewise for the yellow stitches. (See Figures 5, 6, and 7 for evidence that this produces height-two stripes.) One way to accomplish this is to knit $\frac{2k}{5}$ stitches in red, then $\frac{2k}{5}$ stitches

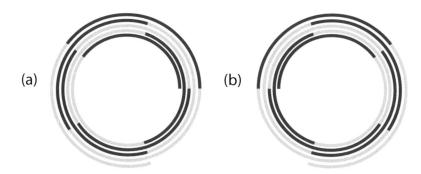

Figure 5. Helix striping that produces height-two stripes of two colors, in (a) left-handed and (b) right-handed versions.

Figure 6. Height-two precessing stripes in two colors, for a left-hander (top) and a right-hander (bottom).

in yellow, knit $\frac{2k}{5}$ stitches in red again and then knit $\frac{2k}{5}$ stitches in yellow again, and continue in this fashion. The first use of red covers stitches 1 to $\frac{2k}{5}$ and the first use of yellow covers stitches $\frac{2k}{5}+1$ to $\frac{4k}{5}$. The second time we knit with red, we'll be knitting from stitch $\frac{4k}{5}+1 \equiv \frac{-k}{5}+1$ mod k to stitch $\frac{k}{5}$ mod k, and thus we overlap our previous red segment by $\frac{k}{5}$ stitches, or half of the segment length. Figure 5 shows this in action. However, it's clear that we'll need more than two strands: when we want to return to red we'll have only knit $\frac{4k}{5}$ stitches and the end of our red yarn strand is at stitch $\frac{2k}{5}$ mod k. Therefore we'll add another strand of red, which will take us to stitch $\frac{k}{5}$ mod k and then we'll need another strand of yellow, which will take us to stitch $\frac{3k}{5}$ mod k, and another strand of red will take us to position 0 mod k (finally!) but we need to knit in yellow for another $\frac{2k}{5}$ stitches to get to the end of the first strand of red. In total, we need six strands, three of each color. (Knitting instructions: knit until arriving at the *second* free end of yarn in the other color.)

This, by the way, creates stripes that *precess*. That is, they travel upwards in the direction opposite that of the knitting, as in Figures 5 and 6. It turns out that stripes that progress can be created by knitting $\frac{2k}{3}$ stitches in each color and using four strands, two of each color.

(See Figure 7.) In addition to changing the number of colors and the (uniform) height of the stripes, there are also other ways to generalize the striping, but we will not address them here (see Section 3.3).

Now, one way to construct n colors of stripes of height h is as follows (though this would correspond to a ridiculous and impractical knitting pattern). If we knit h stitches in each color, this will use nh stitches, and we will obtain stripes if the number of stitches in each round is one more (or one less) than nh. For precessing stripes, let $k = nh + 1$; for progressing stripes, let $k = nh - 1$. In this way, the second use of any color will overlap the first use of that color by $h - 1$ stitches. Thus, one stitch of each color segment will overlap with $h - 1$ other segments of the same color, producing stripes of height h. See Figure 8 for a knitting chart highlighting this overlap in a case where $n = 2, h = 5$, and the stripes progress right-handedly. Let's look at our original example in this light, where $n, h = 2$. Then $k = 5$ for precessing stripes and $k = 3$ for progressing stripes. In either case, we are to knit two stitches per color. As we will see later, any positive multiple of $nh + 1$ (respectively $nh - 1$) would work for k, as well, and generally k will need to be at least 50 stitches for any practical knitting purpose.

Figure 7. Height-two progressing stripes in two colors, for a left-hander (left) and a right-hander (right).

From this, questions arise: Are there other possible constructions to get n colors of stripes of height h? What construction produces the smallest k? To answer this, we will start by making a list of mathematical constraints that are induced by the knitting practicalities of the problem. Then, we will solve the resulting equations.

Suppose we have $\frac{mk}{r}$ stitches for each color and n colors. (A rational multiple of k is needed to produce an integer number of stitches.) Without loss of generality, we may assume that $\gcd(m,r) = 1$. Segments of color that are more than one round long produce horizontal stripes, so in order to retain helical striping $m < r$ so that $\frac{mk}{r} < k$. To obtain stripes of height h, the second occurrence of the first color must overlap the first occurrence of the first color by a fraction of the color segment length, namely $\frac{(h-1)}{h}$. In this way, $\frac{1}{h}^{\text{th}}$ of each color segment will overlap with $h - 1$ other segments of the same color. Figure 8 shows this for an example where the color segments are five stitches each.

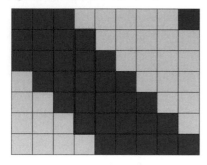

Figure 8. A partial knitting chart showing the overlapping of color segments to produce height-five stripes.

The first use of all colors occupies $\frac{nmk}{r}$ stitches. Together with the above constraints, this means that we need $\frac{(n+1)mk}{r} \equiv \frac{(h-1)mk}{hr}$ mod k for precessing stripes or $\frac{nmk}{r} \equiv \frac{km}{hr}$ mod k for progressing stripes. Furthermore, the second occurrence of the first color must overlap with the first color in the *second* round of knitting—so a stronger condition holds, namely $\frac{nmk}{r} = k - \frac{mk}{hr}$ for precessing stripes (i.e., $nm = \frac{(hr-m)}{h}$) and $\frac{nmk}{r} = k + \frac{mk}{hr}$ for

progressing stripes (i.e., $nm = \frac{(hr+m)}{h}$). Notice that this condition is independent of k, though of course $\frac{mk}{r} \in \mathbb{Z}$ is a necessary constraint.

Let's solve. We can write $nm = \frac{(hr \pm m)}{h}$ and deal with the precessing ($-$) and progressing ($+$) cases simultaneously. Multiplying through by h gives us $nmh = hr \pm m$. This can be rewritten as $nmh \mp m = hr$, so that $m(nh \mp 1) = hr$. Thus m divides hr, but we know that $\gcd(m,r) = 1$ and therefore m divides h. We could instead rewrite $nmh = hr \pm m$ as $\pm m = nmh - hr = h(nm - r)$ so that h must divide m. Therefore, $m = h$.

Returning to our original equation $nm = \frac{(hr \pm m)}{h}$, we may now rewrite this as $nh = \frac{(hr \pm h)}{h}$ or $n = \frac{(r \pm 1)}{h}$, i.e., $r = nh \mp 1$. That separates to $r = nh + 1$ for precessing stripes and $r = nh - 1$ for progressing stripes. Additionally, because $\frac{mk}{r} \in \mathbb{Z}$ and $\gcd(m,r) = 1$, we know that r divides mk but has no common factor with m and thus r must divide k. That is, k must be a multiple of $nh \mp 1$.

Notice that these calculations are consonant with our original example, where $h = n = 2$ and we knit $\frac{2k}{5}$ (respectively $\frac{2k}{3}$) stitches per segment. We see that $m = 2$ and $r = 5$ (respectively $r = 3$), and that k has to be a multiple of 5 (respectively multiple of 3).

From this we can answer our questions: no, there are no other constructions that produce n colors of stripes of height h, and the smallest one is when $k = nh \mp 1$. If we let $k = j(nh \mp 1)$, then we will knit $\frac{mk}{r} = \frac{hk}{r} = \frac{hj(nh \mp 1)}{(nh \mp 1)} = hj$ stitches per color.

We have now proven the following theorem.

Theorem 1 *To produce n colors of stripes of height h using generalized helix striping, the number of stitches per round must be a multiple of $nh + 1$ (for precessing stripes) or $nh - 1$ (for progressing stripes).*

In practice, we need to know the number of strands of yarn of each color that are required. Let the number of stitches in a round be $k = j(nh \pm 1)$ and let s be the number of strands of each color. Consider that using

Figure 9. Precessing stripes (for a left-hander) of height three using four colors.

Figure 10. Progressing stripes (for a left-hander) of height three using four colors.

each color once creates nhj stitches and that after using all s strands of each color (i.e., after knitting $s(nhj)$ stitches), we would want to pick up the first strand of the first color and repeat the whole shebang. For this to work out, we need to use the nth color to knit stitches $0 \mod k$ through $hj \mod k$ because the first strand of the first color was left at stitch hj. Thus we want to solve the equation $s(nhj) \equiv hj \mod k$. This can be rewritten as $s(nhj) \equiv hj \mod j(nh \pm 1)$ and then as $s(nh) \equiv h \mod (nh \pm 1)$. Notice that this equation is independent of the multiple of $(nh \pm 1)$ chosen to create k.

Because n, h are relatively prime to $nh \pm 1$, they have multiplicative inverses. This produces the (not yet adequate) solution $s \equiv n^{-1} \mod (nh \pm 1)$. But what is n^{-1}? It is straightforward to see that $n^{-1} \equiv h \mod (nh - 1)$ because $nh \equiv 1 \mod (nh - 1)$. Likewise, $n^{-1} \equiv -h \mod (nh + 1)$.

We've just proven another theorem.

Theorem 2 *To produce n colors of stripes of height h using generalized helix striping, one will need $(n - 1)h + 1$ strands of yarn in each color (for precessing stripes) or h strands of yarn in each color (for progressing stripes).*

Let's attempt an example, wherein we have four colors and would like precessing stripes of height three (see Figure 9). Therefore $n = 4, h = 3$, and k must be a multiple of $4 \cdot 3 + 1 = 13$. We will choose $k = 65$ (a fine number of stitches for a medium-sized sock), which means that

we will knit with each color for $3 \cdot 5 = 15$ stitches. We will need $(4 - 1)3 + 1 = 10$ strands of yarn in each color. Whew!

For progressing stripes (see Figure 10), we would need k to be a multiple of 11 such as 66, and knit stripe segments of length 18 stitches each. We would need three strands of yarn in each color. Notice that there is a dramatic difference between the number of strands needed for precessing stripes and progressing stripes.

Heuristically, we expect that precessing stripes will require more strands than progressing stripes. The reason for this is that when creating progressing stripes, we'll always need exactly h strands before we can pick up the end of the first strand to reuse it. However, with precessing stripes, after h strands we merely have the end of the second color above the end of the first strand, and have to "wait" until the end of the nth color comes around. This idea is borne out in the mathematics—$(n - 1)h + 1$ will always be larger than h, because to have any sort of striping at all, $n \geq 2$ so that $(n - 1)h + 1 \geq (2 - 1)h + 1 = h + 1 > h$. In practice, this means that progressing stripes always use fewer strands of yarn than precessing stripes. Lesson: knit with progressing stripes. (We will do so in Section 4.)

A related practical concern is the number of vertical rows necessary to carry a given yarn. This will always be the same as the number of strands needed in each color—after all, when one reuses a strand, one carries it up the same number of rows as there are other segments of this color.

Yet another practical concern is whether a progressing-stripe pattern has a corresponding precessing-stripe pattern on the same number of stitches (and therefore has stripe segments of the same stitch length). This would be useful if one were designing a pair of mirror-image socks. Let j_1 be the multiple of $nh + 1$ used for the precessing sock and let j_2 be the multiple of $nh - 1$ used for the progressing sock. Let's see: we would need $k = j_1(nh + 1) = j_2(nh - 1)$ and would knit $j_1 h = j_2 h$ stitches per color. These two equations have no solutions for j_1, j_2; the second requires that $j_1 = j_2$ and the first requires that $j_1 \neq j_2$. Perhaps we can approximate instead, and have k_1 stitches per round of the precessing sock and k_2 stitches per round of the progressing sock. We will require that $| k_1 - k_2 | \leq 2$ so that the change in sizing is unlikely to be noticeable, and we will likewise require that $| j_1 h - j_2 h | \leq 2$ so that the difference in the stripe segment length is unlikely to be noticeable. This second constraint can be rewritten as $| j_1 - j_2 | \leq \frac{2}{h}$, so that for $h \geq 3$ this forces $j_1 = j_2$. We also find that $| k_1 - k_2 | = | j_1(nh + 1) - j_2(nh - 1) | = | (j_1 nh - j_2 nh) + j_1 + j_2 | \leq 2$, so when $j_1 = j_2$, we have $| 2j_1 | \leq 2$. That is, $j_1 = j_2 = 1$, so that $k_1 = nh + 1, k_2 = nh - 1$. Unless n and h are rather large (or the yarn quite bulky), that's not going to produce enough stitches to make a reasonable sock. What about the case of $h \leq 2$? For a moment, let's look at $h = 1$ and $h = 2$ separately. If $h = 1$, then $| k_1 - k_2 | = | j_1(n + 1) - j_2(n - 1) | = | n(j_1 - j_2) + j_1 + j_2 | \leq 2$. If $h = 2$, then $| j_1 - j_2 | \leq 1 | k_1 - k_2 | = | j_1(n2 + 1) - j_2(n2 - 1) | = | 2n(j_1 - j_2) + j_1 + j_2 | \leq 2$. As before, $| j_1 - j_2 | \leq 2$. These inequalities have lots of solutions. For example, if $h = 1, n = 8, j_1 = 8$, and $j_2 = 10$, then $k_1 = 8(8 + 1) = 72$ and $k_2 = 10(8 - 1) = 70$. Or if $h = 2, n = 4, j_1 = 7$, and $j_2 = 9$, then $k_1 = 7(4 \cdot 2 + 1) = 63$ and $k_2 = 9(4 \cdot 2 - 1) = 63$. These are both reasonable numbers of stitches for

socks. (A few other likely sock vectors: (h, n, j_1, j_2) can be $(1, 9, 7, 9), (1, 7, 7, 9), (2, 6, 5, 6), (2, 6, 6, 7)$, or $(2, 4, 6, 8)$.)

2.2 Generalized Helix Striping: The Braid Words

We begin by making a model for the tangling of yarn strands during helix knitting. Imagine that there is a row of balls of yarn pinned in place within a knitting bag, and that yarn travels from each ball to the interior of a knitted cylinder. (We are knitting in the round, after all.) In this model, the yarn gets to the interior of the cylinder through the cast-on opening rather than through the opening occupied by the needle(s), i.e., through the "bottom" of the work rather than over the "top" of the work.

Our model will use braid words to describe the entangled strands of yarn. As the knitting progresses, the strand that forms new stitches passes over strands of yarn not in use. When colors are changed, strands are twisted together in order to avoid holes in the finished work. The braid letters are formed near the work and collect nearer to the balls of yarn, so we can think of the knitter looking at the braid as holding the knitted cylinder in her lap and examining the braid as it passes into her knitting bag. The visual top of the braid is near the balls of yarn and holds the oldest tangling, and the visual bottom of the braid is near the knitted cylinder (probably inside it in reality, but we can push it just outside) and holds the most recent tangling.

This informs the notation we will use to create and interpret braid diagrams. In contrast to our previous work [2] but in accordance with a standard braid notation [1], we label strands from left to right, read crossings from top to bottom, and consider σ_i to be the movement of the ith strand over the $(i+1)$st strand. That means that σ_i crosses a left strand over a right strand, and σ_i^{-1} crosses a right strand over a left strand. These are shown in Figure 11.

Figure 11. At left, σ; at right, σ^{-1}.

Here are the questions that now arise: What happens when a knitter changes colors? What happens when she completes a round? What happens when she knits past a strand (as required for helix striping with stripes of height larger than one)? We deal with these one at a time.

First, when a knitter changes colors, she places the first color horizontally across the work (in the "forward" direction), and then picks up the strand of the next color to knit with it. This prevents a hole from developing, as when the two strands are twisted in this fashion, they pull toward each other (rather than away from each other as they would if not twisted). If the first color is emanating from ball i and the second color is emanating from ball $i + 1$, then this produces σ_i^2. The reason is that as the knitter places the first color horizontally across the work, strand i passes across strand $i + 1$ a few inches away from the work. This places the first color in the $i+1$st position and the second color in the ith position in the braid. Then, when the knitter picks up the second color to knit with it, she will pull it "forward" as she knits and this passes strand i across strand $i + 1$ a few inches away from the work. This explanation is reflective of left-handed knitting. A right-handed knitter would knit with her first color emanating from ball i and twist this strand across the second color, emanating from ball $i - 1$; this would produce σ_{i-1}^{-2}. For ease of notation, we will use left-handed knitting for the remainder of the analysis.

Similarly, when a knitter knits past a strand, as when she knits with strand i in one color, sees next strand $i + 1$ in the same color, but wants to knit with strand $i + 2$ in a second color, this will produce σ_i as strand i will pass across strand $i + 1$ a few inches away from the work.

The matter of completing a round is a bit more complicated, as we will see from the following example. Let us line up the strands according to the initial position of the balls of yarn. For the sake of clarity, we'll use a different color for each strand in the diagram. When we switch colors, we'll insert a σ_i^2 and when we pass a strand we'll insert a σ_i. Suppose we are knitting along, switch colors once, pass a strand, and then switch colors again as shown in Figure 12.

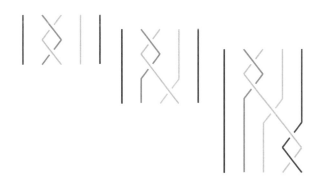

Figure 12. The three images show the twists resulting in five strands of yarn after (left) switching colors once, (center) switching colors and then passing an inactive strand, and (right) switching colors, passing an inactive strand, and switching colors a second time.

Now what? If we want to switch from the purple strand to the red strand, we somehow have to have the purple appear to the left of the red on the diagram. How do we do that? We could have the purple cross all the other strands, so that it's on the left. But should it go over or under all the strands? Or, we could bring the red strand across so it's all the way to the right. Images of these three modeling options appear in Figure 13. So far, there is no indication that any of these options is preferable to the others.

Consider what is physically happening when knitting in the round takes place. The balls of yarn are fixed and the knitting rotates relative to them. In left-handed knitting, the knitting rotates clockwise. In slow motion, it appears that the left-most strand passes over all of the

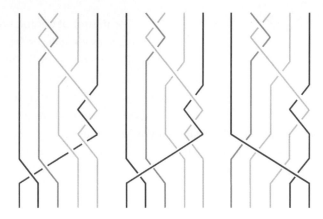

Figure 13. Three options for achieving the goal of having the purple strand appear to the left of the red strand.

other strands when it rotates around the back of the cylinder and appears on the knitter's right. However, this does not solve the problem. Suppose a bunch of strand-twisting is happening in the front of the cylinder, closest to the knitter; between which two braid letters should the strand pass-over maneuver be inserted? It turns out that it doesn't matter! If we rotate the knitter instead of the knitting, we may thereby engineer each strand pass-over to be placed in any of several spots. We may as well model the situation by adding in a complete rotation of the strands at the end of each round, shown in Figure 14. If there are n strands, this corresponds to appending $(\sigma_1\sigma_2 \dots \sigma_{n-1})^{n-1}$ to the braid word. In practice, this also means that we can minimize the tangling of the strands by un-rotating the whole work once per round. Luckily, knitting some samples and checking the accompanying braids bear this out in practice.

Notice that each braid letter used so far—for switching colors, passing over a strand, or completing a round—is a σ_i and none are σ_i^{-1}. For ease of reading braid words on the page, we will henceforth write the subscripts only and not the σs. (Because there are no inverse elements, no confusion should arise.) For example, instead of writing $\sigma_1^2\sigma_2\sigma_3^2$, we write $1^2 2 3^2$.

Figure 14. The rotation of strands shown in this diagram reflects the completion of a round of helix striping.

We return to our original question: what braid words are generated in generalized helix striping? We will again start with the basic construction, pictured in Figure 15, using three colors and stripes of height one. As a braid word, we obtain 11221212. Interestingly, using the usual braid relations, $i(i + 1)i = (i + 1)i(i + 1)$ and $i(i+j) = (i+j)i$ when $j > 1$ (see [1] for details), this turns out to be equivalent to $(12)^4$ (really, $(11221212)^3 \equiv (12)^{12}$).

Figure 15. The three-strand generating braid word $11221212 \equiv (12)^4$.

The practical implication of this mathematics is that when doing the ordinary helix striping with three colors, one can basically dangle the work and the braid from the strands emanating from the balls of yarn, and the work should spin around and the braid should untangle itself. (In reality, the balls move around in the knitting bag and reduce the efficacy of this tactic.)

Does this property generalize? For $(n + 1)$ colors, we obtain the generating word $1122\ldots nn(12\ldots n)^n$. (An image of this word for $(n + 1) = 4$ is shown in Figure 16.)

Figure 16. With four colors, the generating braid word is $112233(123)^3$.

The braid word relations imply that $12\ldots nj = (j+1)12\ldots n$ for $j < n$, which allows a lot of rewriting but nothing so significant that we obtain a simpler braid word. Basically, the term $12\ldots n$ can be moved within a round of knitting, but not past the end of the round (this corresponds to our slow-motion rotation strand pass-over observation earlier); it seems more convenient to keep all $(12\ldots n)^n$ in one place as it physically corresponds to a twist of the whole work. The only practical conclusion seems to be that we might as well try dangling the work and braid from the balls of yarn and letting the braid untwist in various spots, but we would then still have to deal with the remaining pile of i^2 strand twists.

We now consider the generalization to heights other than one. Again, we start with the simplest example, of two colors and height-two stripes. For progressing stripes, we have four strands and the generating word $11233(123)^3 1223(123)^3$. For precessing stripes, we have six strands and the generating word $11233455(12345)^5 1223445(12345)^5$. These are both rather ugly, and no amount of braid-word relation manipulation appears to simplify either of them.

By using the instructions given in Section 4.3 for how to knit generalized helix stripes of any height with any number of colors, we can derive the generalized generating braid word and see that it depends on the number of colors n and the number of strands s used for each color (rather than on n and the height h). It turns out to be

$$1^2 2 \ldots s(s + 1)^2(s + 2) \ldots (2s)(2s + 1)^2(2s + 2) \ldots ((n - 1)s)$$
$$((n - 1)s + 1)^2((n - 1)s + 2) \ldots (ns - 1)(12\ldots(ns - 1))^{ns-1}$$
$$12^2 3 \ldots (s + 1)(s + 2)^2(s + 3) \ldots (2s + 1)(2s + 2)^2(2s + 3) \ldots$$
$$((n - 1)s + 1)((n - 1)s + 2)^2((n - 1)s + 3) \ldots (ns - 1)(12 \ldots$$
$$(ns - 1))^{ns-1} 123^2 \ldots (s + 2)(s + 3)^2(s + 4) \ldots (2s + 2)(2s + 3)^2$$
$$(2s+4) \ldots ((n - 1)s + 2)((n - 1)s + 3)^2((n - 1)s + 4) \ldots (ns - 1)$$
$$\ldots (12 \ldots (ns-1))^{ns-1} 12 \ldots (s - 1)s^2(s+1) \ldots (2s-1)(2s)^2(2s+1)$$
$$\ldots ((n - 1)s - 1)((n - 1)s)^2((n - 1)s + 1) \ldots (ns - 1).$$

There is something interesting to be seen from graphical representations of these yucky generators, though: as shown in Figure 17 (for three strands each of two colors), the strands of a single color do not tangle together, but only with strands of other colors. (This makes sense from a knitting perspective, as the strands of a single color do not interact with each other.) The diagram shows progressing stripes, but the situation is similar for precessing stripes. Only global twisting (i.e., $(12 \ldots s)^k$) of the strands of a single color happens.

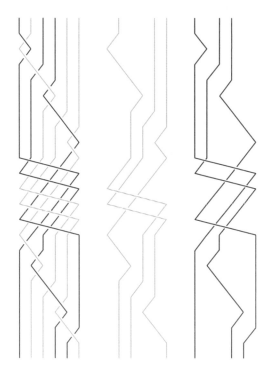

Figure 17. Check out the lack of intracolor entanglement!

3   Teaching Ideas

The ideas in number theory are so universal in mathematics education that we often study them (or teach them) without even realizing that this is the subdiscipline under discussion. Luckily, this means that most levels of education can use the mathematics in Section 2.1! Teaching ideas for braids are given in Section 3 of [2].

3.1   Upper Elementary and Middle Grades

Students who are learning about divisibility can start to think about *vertical* striping. Modeling the object as knitted in the round will allow for a generalization to diagonal striping later. Note that when knitting in the round not every yarn end from the previous row will be available for knitting at the appropriate stitch on the next row. Thus, a vertically striped object can't be knitted in the round without using very thin stripes and employing intarsia methods, or using a separate ball of yarn for each stripe of each row.

To link divisibility and knitting, the situation to set before the students is that they are to create stripes of equal width. Given a certain number of stitches, what are the possibilities for choosing a number of colors and stitch width of stripes? For example, if there are 21 stitches, there can be 3 colors of stripes with 7 stitches each, or 7 colors of stripes with 3 stitches each. It is likely that students will also suggest using fewer colors but alternating them (perhaps unevenly). Using a number of stitches such as 60 will produce many more answers (including 5 stripes of 12 stitches each, 10 stripes of 6 stitches each, 20 stripes of 3 stitches each, and 4 stripes of 15 stitches each). This can be related to the construction of factor trees.

For students just learning multiplication, this type of question can be phrased as giving a number of stripes and a stitch width and asking what total number of stitches is needed for the project.

Once the ideas in vertical striping have been mastered, the question can be modified to ask what happens when the number of stitches changes by one. In practice, a teacher would choose the total number of stitches and ask the class to agree on a corresponding number of stripes and width of stripes. Here we will use variables so as to keep the problem applicable to any classroom. If the number of stripes is n and each stripe is w stitches wide (that is, we knit w stitches in each color), what happens if the number of stitches is $nw + 1$ and what happens if it is $nw - 1$? In order to envision the situation, students can be asked to imagine a tube with nw stitches. This tube is then dyed or painted with n stripes, each w stitches wide; then, the tube is unraveled and reknitted using $nw + 1$ or $nw - 1$ stitches. (A physical model could be made, but the unraveling and reknitting would take quite a lot of time.) The answer is that stripes will precess (with $nw+1$ stitches) or progress (with $nw-1$ stitches).

This problem can be generalized slightly to consider $nw + 2$ or $nw - 2$ total stitches, and then $nw + k$ or $nw - k$ total stitches. Students can first recognize that the type of stripe will not change, and then notice that the apparent slope of the stripes will decrease as k increases. This is a good opportunity to discuss the ideas underlying slope. For example, steeper slopes have greater vertical change than horizontal change. The vertical change can be approximated by the many overlapping rows of stitches stacked atop each other within a stripe, and the horizontal change can be approximated by the width w of each stripe. Similarly, more gradual slopes have greater horizontal change than vertical change, as can be seen by the comparatively small number of overlapping rows of stitches stacked atop each other within a stripe.

3.2 High-School and Collegiate Number Theory Courses

Helix striped knitting patterns are a good application of modular arithmetic.

* Students can be given the knitting instructions for height-one helix stripes in three colors and asked to rephrase them in terms of modular arithmetic: which stitches, mod k, are knitted in which colors? Hopefully students will recognize that the answer depends on which row is being knit, and produce a list of answers.

* The same question can be asked for helix stripes in n colors.

* For the n color helix stripe pattern, how many different rows of modular arithmetic instructions are there before repeating? What is the formula for this number in terms of k and n?

* If there are more colors, what properties should n have in terms of k?

* Will colors always be switched at stitch 0 mod k?

* Given n colors, at how many places mod k will colors switch?

Students can also use modular arithmetic to analyze stripes of height two, and then of height h. Without worrying about a value for k, students should think about at what positions, relative to 0 mod k, colors will switch. They should notice that the fraction denominators must be relatively prime to both n and h. This can be the starting point for a discussion of which numbers produce as multiples all residue classes mod m, or equivalently, of the fact that generators for $\mathbb{Z}_m$ are exactly those elements that are relatively prime to m.

The process of setting up and solving the equations given on pages 33–34 can be worked through with students after they have learned how to make divisibility arguments. Theorem 2 can be set up and proven once students have learned that common factors can be cancelled from modular equations and how to compute multiplicative inverses when they exist.

3.3 Variants and Project Ideas

The ideas presented in this subsection are intentionally presented without detail, so as to leave them as open-ended as possible.

★ Generalize to stripes of nonuniform height. For example, instead of having three colors and stripe height four, use three colors with one color having stripe height two, one color having stripe height three, and one color having stripe height four. Or, have five colors, two with stripe height two and two with stripe height four, with the remaining color having stripe height one.

★ Another way of generalizing helix striping is to use the same construction as for rounds of k stitches and height-one stripes with n strands of yarn, but instead of using n different colors, use $\frac{n}{r}$ colors with each color used r strands contiguously. For example, suppose $n = 12$, $r = 3$. Then the strands used will be yellow, yellow, yellow, red, red, red, orange, orange, orange, brown, brown, brown. What height stripes does this construction produce? Perform a similar analysis as in Section 2.1 to see what constraints arise on the variables. From a knitting perspective, is this construction easier or more difficult than the construction given in Section 2.1? In what ways? Is this construction more or less flexible than that given in Section 2.1? How do the cases of progressing and precessing stripes compare? In terms of the braid created by yarn tangling, does this construction induce or avoid intracolor and/or intercolor entanglement?

★ The example in Figure 17 of separating colors in the braid diagram was made for progressing stripes. Here, the only intracolor interaction is caused by completing knitting rounds. Precessing stripes have some additional intracolor interaction. What is this extra spiralling and what causes it? Can the additional spiralling be removed by letting the work hang, or do the spirals of various colors interfere with each other?

4 Crafting a Helix Striped Spiral Nightcap and Bed Socks

The inspiration for this project was a cold head complaint from someone sleeping in a bed placed close to a window in a bedroom far from the woodstove that heats the author's house. The resulting hat is shown in Figure 18.

Matching socks can be made to use up the yarn remaining from the hat, and a sample pair is shown in Figure 19.

Materials

★ Four colors of Valley Yarns Superwash (100% wool, 50g/97 yds.) or Di.Ve' Zenith (100% wool, 50g/120 yds.), one ball each (∼150 yds. total for the hat).

If also making the socks, buy an additional ball, perhaps in a fifth color, for the heels, toes, and cuff (∼350 yds. total for the socks). The smoother the yarn used, the less the frustration caused by frequent tangling.

★ Size 7 needles—circular are recommended for use with magic loop technique, but the use of double-pointed needles is possible.

★ One locking stitch marker (for hat).

★ Two stitch markers (for socks) and 12 locking stitch markers (for Japanese short row heels).

★ (Optional) 16 locking yarn bobbins such as Susan Bates Yarn Bobs.

Figure 18. A helix striped spiral nightcap.

Gauge

★ Five stitches per inch and seven rows per inch.

Abbreviations

★ K (or k) means knit.
★ P (or p) means purl.
★ K2tog (or k2tog) means knit two together.
★ SSK means slip, slip, knit.

Notes

★ The construction of the hat and socks is a helix; that is, it takes advantage of the spiral nature of knitting in the round. Knitting the sequence of colors once causes more than one round to be knitted.

★ When switching colors, place the working strand of the current color horizontally behind the work, so that picking up the free strand of the new color has the new color cross over the old color. This is important so that there will not be holes in the finished item.

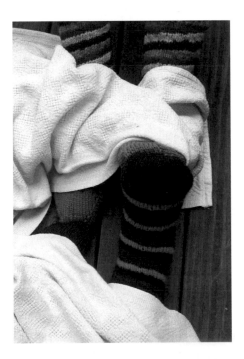

Figure 19. Helix striped spiral bed socks.

★ For both hat and socks, weave in ends during the first rows by carrying the ends with the work (of the appropriate color). Otherwise it's more challenging than usual to weave in the ends separately. One way to avoid unintentional holes when introducing a new ball of yarn is to carry the end of the old yarn across the back under the working yarn for one stitch.

★ As the hat narrows at the top, it becomes difficult to distinguish between nearby strands of yarn. If one is using bobbins, then these can be labeled (e.g., as A, B, C, D) to assist with this difficulty.

★ The socks are not amenable to being done 2-at-a-time as this would result in 32 balls of yarn dangling from the work.

★ There are several different techniques for doing short rows for the socks' heels. All have in common that there is some additional action taken before reversing the direction of knitting. One can use:

- *Wrap-and-turn*: Slip a stitch, move the yarn to the front of the work, slip that stitch back, and move the yarn to the back of the work. Then, turn the work around to purl.

- *Japanese short rows*: Knit an additional stitch, slip the knit stitch back, turn the work around to purl, and place a locking stitch marker on the yarn between the slipped-back stitch and the next stitch to be worked.

- *Priscilla Wild's no-fuss no-muss short rows*: Simply turn the work and slip instead of creating the first worked stitch.

Additionally, an ambidextrous knitter usually does not turn the work to purl, but simply knits in the opposite direction. To accommodate all varieties of short rowing, we will write *reverse direction* as a signal that the knitter should do whatever she prefers for her short rows at that point.

Just as there are different techniques for creating short rows, there are different techniques for hiding the join between the short rows and the rest of the sock. For the wrap-and-turn technique, pick up the wrap and work it together with the wrapped stitch. For Japanese short rows, work the loop of yarn encircled by the locking stitch marker together with the stitch that is 1.5 stitches past the loop. For Priscilla Wild's no-fuss no-muss short rows, knit the next two stitches together, then make 1 from the strand between stitches. This operation will be abbreviated as *close the gap*.

★ There are lots of ways to decrease, and depending on how you knit (standard vs. combined vs. left-handed vs. ...) your k2tog may lean in the same direction that you knit or in the opposite direction that you knit. Likewise, your SSK may lean toward your dominant hand (against the direction you knit) or in the direction of knitting. Thus, we will abbreviate decreases as *decrease forward* to indicate a decrease that leans in the direction of knitting and *decrease backward* to indicate a decrease that leans against the direction of knitting.

Preparations

Prepare by making each of the four balls of yarn into four equally-sized smaller balls of yarn. Winding these 16 balls of yarn onto locking yarn bobbins will reduce the frustration associated with detangling.

We will be using four colors and creating progressing stripes of height 4, so we will need to maintain a multiple of $4 \cdot 4 - 1 = 15$ stitches. (Recall that because the stripe height is greater than two, we cannot create mostly-matching mirror-image striping for the socks.) The hat will begin with $6 \cdot 15 = 90$ stitches and the socks will begin with $3 \cdot 15 = 45$ stitches. The stripes will use $6 \cdot 4 = 24$ and $3 \cdot 4 = 12$ stitches each, respectively.

4.1 Knit the Hat

Render the Ribbing

Cast on 24 stitches in color 1, 24 stitches in color 2, 24 stitches in color 3, and 18 stitches in color 4.

Join to knit in the round, place a stitch marker at the beginning of the round, and k3, p2, k1 in color 4.

Tiny overview. The goal is now to work in pattern (k3, p2) while changing colors every 24 stitches; the next strand of the next color will be directly below the next stitch when it is time to switch. Note that the last six stitches of each color are knitted onto stitches of a different color, i.e., one need only count six stitches when the color of the loops one is knitting changes.

Using a new ball each time, work 24 stitches in color 1 (this is k2, p2, (k3, p2) four times), then color 2 (this is (k3, p2) four times, then k3, p1), then color 3 (this is p1, (k3, p2) four times, k3), then color 4 (this is p2, (k3, p2) four times, k2), and continue adding new balls in color and stitch sequence so that all 16 balls of yarn are incorporated into the hat.

The strand of yarn from the first ball of color 1 should be directly below. Pick it up and work 24 stitches in pattern. Now the strand of yarn from the first ball of color 2 should be directly below. Pick it up and work 24 stitches in pattern.

Continue in this fashion until 12 rounds (not color sequences) or 1.5" have been completed.

Synthesize the Sides of the Hat

Begin knitting all stitches, and continue to change colors every 24 stitches. Continue until the total knitted length is approximately 5" and you are ready to start a round with color 1. (The precise length will depend on your row gauge. Do note that color 1 does not always begin a round.)

Create the Crown

Beginning with color 1 and working in color sequence, decrease 15 stitches by working (k4 k2tog) 15 times over one round, leaving 20 stitches of each color (75 stitches left in a round). Specifically, work (k4 k2tog) four times for colors 1–3, and work (k4 k2tog) three times for color 4; the next round will begin with five stitches of color 4.

Continue knitting for six more rows or almost 1", switching colors after every 20 stitches. Now there are five stitches knitted on top of the next color rather than six.

Beginning with color 3 and working in color sequence, decrease 15 stitches by working (k3 k2tog) 15 times over one round, leaving 16 stitches of each color (60 stitches left in a round). The next round will start with four stitches of color 2.

Continue knitting for six more rows or almost 1", switching colors after every 16 stitches. Now there are four stitches knitted on top of the next color rather than five.

Beginning with color 1 and working in color sequence, decrease 15 stitches by working (k2 k2tog) 15 times over one round, leaving 12 stitches of each color (45 stitches left in a round). The next round will start with three stitches of color 4.

Continue knitting for six more rows or almost 1", switching colors after every 12 stitches. Now there are three stitches knitted on top of the next color rather than four.

Beginning with color 3 and working in color sequence, decrease 15 stitches by working (k1 k2tog) 15 times over one round, leaving eight stitches of each color (30 stitches left in a round). The next round will start with two stitches of color 2.

Continue knitting for six more rows or almost 1", switching colors after every eight stitches. Now there are two stitches knitted on top of the next color rather than three.

Figure 20. The finished spiral (left) and an inverted hat (right).

Beginning with color 1 and working in color sequence, decrease 15 stitches by working k2tog 15 times over one round, leaving four stitches of each color (15 stitches left in a round). The next round will start with one stitch of color 4.

Knit three more rounds, changing colors after every four stitches, where only one stitch is knitted atop the next color.

Beginning with color 3 and working in color sequence, k2tog until there is one stitch of each color remaining.

Finish

Pull a free end of yarn through the stitches, and pull that strand inside the hat. Admire the resulting spiral, shown in Figure 20 (left). Weave in all 16 ends (or 32 ends if you have some left from the set-up rounds).

Alternate endings. Weave in the free end, and pull all ends through to the right side of the hat; trim to form a "tail." Some knitters may prefer to leave the ends on

the interior of the hat and wear it inside-out as shown in Figure 20(right).

Congratulations, you're done!

4.2   Knit the Socks

A finished sock is shown in Figure 21.

Render the Ribbing

Using an invisible cast-on (such as the loop cast-ons given in [5, pp. 65–66]), cast on 44 stitches using the fifth ball of yarn. Place a stitch marker and join to knit in the round. Work eight rounds of 1×1 ribbing. Make 1 (just before marker) for a total of 45 stitches.

Lay Down the Leg

Using a new ball each time, k12 in color 1, then color 2, then color 3, then color 4, and repeat in color sequence a total of four times so that all 16 balls of yarn are incorporated into the sock.

The strand of yarn from the first ball of color 1 should be directly below. Pick it up and k12. Now, the strand of yarn from the first ball of color 2 should be directly below. Pick it up and k12. Continue to knit while changing colors every 12 stitches; the next strand of the next color will be directly below the next stitch when it is time to switch.

Continue striping along until the total knitted length is $7\frac{1}{2}$" or 8" and you are ready to start knitting with color 1. (It does not matter where in the round you are.) Tink (unknit) the last stitch of color 4.

Figure 21. A helix striped sock, modeled.

Hatch the Heel

Prepare for short rows by checking the sixth Note (page 44). That said, switch to the additional ball of yarn (color 5) and reverse direction.

Work 22 stitches, carrying along each free end of yarn available below for 3 stitches to secure it, and reverse direction. Clip the secured yarn and remove the balls for use after the heel is complete.

Work 21 stitches and reverse direction.
Work 20 stitches and reverse direction.
Work 19 stitches and reverse direction.
Work 18 stitches and reverse direction.
Work 17 stitches and reverse direction.
Work 16 stitches and reverse direction.
Work 15 stitches and reverse direction.

Work 14 stitches and reverse direction.
Work 13 stitches and reverse direction.
Work 12 stitches and reverse direction.
Work 12 stitches, close the gap, and reverse direction. (Note that for Japanese short rows, the stitch slipped to reverse direction is the one created when closing the gap.)
Work 13 stitches, close the gap, and reverse direction.
Work 14 stitches, close the next two gaps (one older, from the decreasing short rows, will be below the newer one created two rows prior), and reverse direction.
Work 15 stitches, close the next two gaps, and reverse direction.
Work 16 stitches, close the next two gaps, and reverse direction.
Work 17 stitches, close the next two gaps, and reverse direction.
Work 18 stitches, close the next two gaps, and reverse direction.
Work 19 stitches, close the next two gaps, and reverse direction.
Work 20 stitches, close the next two gaps, and reverse direction.
Work 21 stitches, close the next two gaps, place marker, and reverse direction. This marker (as does the next marker placed) separates the stitches for the top of the foot from those of the bottom of the foot for later use in toe shaping.
Work 22 stitches, place marker, and close the next two gaps using color 1; the second gap is in color 5's free end. Now continue to complete 12 stitches in color 1. K8 in color 2, and close the remaining two gaps.

Form the Foot

K3 in color 2. Add in a ball of color 3 and k12. Add in a ball of color 4 and k12. Continue striping along as before the heel, adding back in balls of yarn as you come to the places where you expect them to be.

Stripe along until you are ready to begin the toe (3″ from the tip of the toe to customize, or $4\frac{1}{2}$″ from the heel for a medium foot). (If for some reason one color runs out, as might happen when taking a chance with partial balls of yarn, just go monochrome for the remainder of the sock.)

Turn Out the Toe

Switch back to color 5 and carry along each free end of yarn available below for three stitches to secure it. In this fashion, knit a bit more than one row, and then knit to two stitches before the nearest marker.

See the last Note (page 44) for notation and (decrease backward, k1, decrease forward, knit to two stitches before second marker, decrease backward, k1, decrease forward, complete round, knit two rounds) six times.

Decrease in your favorite direction, and Kitchener (graft) the remaining stitches. (Be sure your stitches are aligned so that the graft is oriented parallel to the tips of your toes.)

Synthesize a Second Sock

Work second sock as described above. To create a sock with stripes spiralling in the opposite direction, substitute purls for knits and carry the yarns on the outside of the sock. (Make sure that the short row wraps/loops are hidden on the purl side of the sock.) Turn the sock inside-out when complete. Look at your socks!

4.3 How to Knit Generalized Helix Stripes, in General

Let's review. In Section 2, we gave instructions for constructing helical stripes of height one. In short, these are the instructions.

For Three Colors

* Cast on $\frac{k}{2}$ stitches in colors 1, 2; join to knit in the round.

* Knit in color 3 ($\frac{k}{2}$ stitches) until reaching the free end of color 1; lay the free end of color 3 across the work, pick up the free end of color 1 so it crosses over color 3, and knit in color 1.

* Generically, knit in the present color until the next free end of yarn, pick up that free end, and knit in the new color.

* Repeat until work is desired length.

For *n* Colors

* Cast on $\frac{k}{(n-1)}$ stitches in colors $1, 2, \dots, n-1$; join to knit in the round.

* Knit in color n ($\frac{k}{(n-1)}$ stitches) until eaching the free end of color 1; lay the free end of color n across the work, pick up the free end of color 1 so it crosses over color n, and knit in color 1.

* Generically, knit in the present color until the next free end of yarn, pick up that free end, and knit in the new color.

* Repeat until work is desired length.

Here are the instructions for knitting helical stripes of height h using n colors.

For *n* Colors, Progressing Stripes of Height *h*

* Cast on $\frac{nhk}{(nh-1)}$ stitches in colors $1, 2, \dots, n-1$ and $\frac{nhk}{(nh-1)} - \frac{nk}{(nh-1)}$ stitches in color n.

* Join to knit in the round.

* Knit $\frac{nk}{(nh-1)}$ stitches in color n, then $\frac{nhk}{(nh-1)}$ stitches in colors $1, 2, \dots, n$ using a second, then third, ..., then n^{th} strand of each.

* At this point the free end of the first strand of color 1 will be available; lay the free end of color n across the work, pick up the free end of color 1 so it crosses over color n, and knit in color 1 until the free end of the first strand of color 2 is available. (This is also the first time *any* free end in color 2 will be available.)

* Generically, knit color j mod n until the next free end of yarn in color $j+1$ mod n is available, pick up that free end, and knit.

Notice that each time, one passes over $h-1$ strands of color j mod n on the way to the next free end of yarn in color $j+1$ mod n.

For n Colors, Precessing Stripes of Height h

* Cast on $\frac{nhk}{(nh+1)}$ stitches in colors $1, 2, \ldots, n$ and $\frac{nk}{(nh+1)}$ stitches in color 1 (using a second strand).

* Join to knit in the round.

* Knit $\frac{nhk}{(nh+1)} - \frac{nk}{(nh+1)}$ stitches in color 1, then $\frac{nhk}{(nh+1)}$ st in colors $2, \ldots, n$ using a second strand of each. Continue knitting using a third, then ... s^{th} strand of each.

* At this point the free end of the original strand of color 1 will be available; lay the free end of color n across the work, pick up the free end of color 1 so it crosses over color n, and knit in color 1, passing $s-1$ strands of color 2, until the first strand used of color 2 is available.

* Generically, knit color j mod n until the sth free end of yarn in color $j+1$ mod n is available, pick up that free end, and knit.

Notice that each time, one passes over $s-1$ strands of color $j+1$ mod n on the way to the lowest free end of yarn in color $j+1$ mod n.

Bibliography

[1] Adams, Colin. *The Knot Book.* W. H. Freeman, New York, 1994.

[2] belcastro, sarah-marie, Szczepański, Amy, and Yackel, Carolyn. (K)not Cables, Braids, In *Making Mathematics with Needlework*, edited by sarah-marie belcastro and Carolyn Yackel, pp. 119–134. A K Peters, Ltd., Wellesley, MA, 2008.

[3] Cairn, Charisa Martin. "No Muss No Fuss No Wraps No Holes Short Row Heel." Available at http://pulsh.blogspot.com/2008/07/video-demo-of-no-muss-no-fuss-no-wraps.html, July 8, 2008.

[4] Hamer, Joan. "Helix Striped Cap." Available at http://www.knitlist.com/97gift/helixcap.htm and http://woolworks.org/patterns/helixcap.txt, 1997.

[5] Stanley, Montse. *Knitter's Handbook : A Comprehensive Guide to the Principles and Techniques of Hand-knitting*. Reader's Digest, Pleasantville NY, 1993.

CHAPTER 3

a knitted cross-cap

EMILY PETERS

1 The Math and Motivation behind the Pattern

This project arose when I realized an easy way to do self-intersecting knitting, described in Section 2. Self-intersecting knitting is necessary for the depiction of some mathematical objects, including the cross-cap immersion of the real projective plane. The real projective plane, $\mathbb{RP}^2$, is the space of all lines through the origin in $\mathbb{R}^3$, with a distance defined by the (smaller) angle between two given lines. Each line passes through a unit sphere at two antipodal (opposite) points. Topologically, the real projective plane is equivalent to identifying (i.e., gluing together) every pair of antipodal points on a sphere. $\mathbb{RP}^2$ is a nonorientable surface, and cannot be embedded in $\mathbb{R}^3$.

The word *cross-cap* refers to a particular image of $\mathbb{RP}^2$ in $\mathbb{R}^3$. We construct it by starting with a sphere decomposed into three parts, as in Figure 1(b): a southern hemisphere, a northern hemisphere, and an equator.

Identifying antipodal points merges the southern and northern hemispheres, and gluing the equator onto the resulting boundary produces a disk. Figure 1(c) shows the disk, with the inside and outside colored differently. To finish the required identification, we must identify antipodal points along the equator. For convenience, we shape the equator into a figure 8 (the left front half of the disk is pushed through and past the left back half of the disk) as shown in Figure 1(d), where the coloring indicates how the sheets flow through the induced intersection. We then stretch the twisted disk as in Figure 1(e) so as to identify antipodal points on the equator; this is shown in Figure 1(f), where the coloring has changed to show that there are no longer two "sides" to the surface.

Notice that in Figure 1(d)–(f), the self-intersection is evidently an intersection, and not a pinch, because of the difference in color between its "inside" and "outside." With knitting, we can achieve the same effect using the contrast between stockinette and reverse stockinette stitch. But because $\mathbb{RP}^2$ is nonorientable, we need

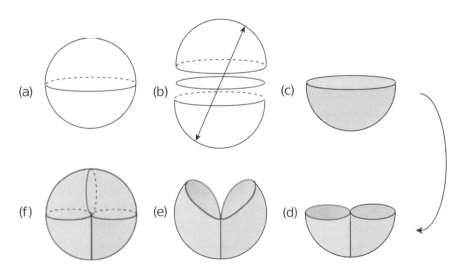

Figure 1. Constructing the cross-cap.

Figure 2. A finished knitted cross-cap.

to fade from stockinette, through moss stitch, and into reverse stockinette when we attach the two halves of the equator to each other.

2   Crafting the Knitted Cross-Cap

Magic loop is one of my favorite knitting techniques; a single set of (long!) circular needles can make a tube of any smaller diameter, eliminating the need to use double-pointed needles or have circular needles of varying lengths. In this project, pictured in Figure 2, magic loop knitting makes it easy to knit a self-intersecting surface.

Materials

* One skein of Noro Kureyon (100% wool, 50g, 110 yds.) is more than enough. (Any worsted weight wool can be substituted.)

* ∼2 meters of scrap yarn for provisional cast-on.

* Size 6 (4mm) 30″ (or longer) flexible circular needles.

* Four stitch markers, one of which is distinctive.

* 25–50 grams of polyester fiberfill.

* Yarn needle, tapestry needle, or crochet hook for finishing.

* Optional: locking stitch marker or safety pin.

Gauge

Gauge is essentially unimportant, but what is important is to knit fairly tightly so that the stuffing does not show through when the cross-cap is finished. If substituting yarn, reduce needle size from that suggested on the ball band.

Abbreviations

* K (or k) means knit.

* P (or p) means purl.

* K2tog (or k2tog) means knit two together.

* SSK means slip, slip, knit these two together.

* P2tog tbl (or p2tog tbl) means purl two together through the back loop.

* P2tog (or p2tog) means purl two together.

Figure 3. Magic loop knitting a cross-cap. Notice that the 30" needle is just barely long enough for the job.

Magic Loop Technique

Magic loop knitting is a technique for knitting pieces with a relatively small diameter in the round, using a very long, very flexible circular needle. If your needle is too short (say, less than 12" longer than the circumference of your object), or not flexible enough, you'll feel as if you're constantly wrestling to keep the needles in place.

The idea is to convert the round nature of the knitting back to two needle knitting by dividing the stitches into two relatively equal groups and knit each group alternatively on the two stiff needles while the other half of the stitches remain out of the way on part of the flexible cable. The excess cable is pulled out in loops between the two groups of stitches.

In more detail, when you get to the end of one group of stitches (call it group A), the passive stiff needle is free and the active needle has all of group A's stitches on it. Pull the active needle forward so that the group A stitches are also on the cable. Now pull the passive needle backwards, so the group B stitches slide from the cable onto the needle. Free up a loop of cable behind the active needle, and then begin knitting the group B stitches.

Because the cable is long and flexible, adjacent stitches in the round can be treated as though they are in separate groups without inducing gaps in the knitting, as long as one is careful. See Figure 3.

For more information, see [1]. In the second half of this pattern, we use the magic loop technique to cause the fabric to self-intersect.

A note on decreasing stitches. We recommend k2tog and SSK to achieve right- and left-leaning knit decreases, and p2tog tbl and p2tog to achieve right- and left-leaning purl decreases, but you may substitute your favorite decreases if desired. For more detailed information, see [2]. (Note that because we are knitting in the round, the purl decreases may seem to slant in the direction opposite to that described in knitting books such as [3].)

Instructions

The big picture. This project is knit in two parts. First, the disk is knit from the boundary to the apex (much like a hat), and then stitches are picked up from this boundary in a fancy way, like a figure 8, so as to finish by knitting the self-intersecting portion of the surface to its apex (much like a self-intersecting and therefore impractical hat).

Start the cross-cap. We'll knit the nonintersecting half of the cross-cap first. Because Noro Kureyon is a self-striping yarn, and we're starting in the middle of the project, we want to start knitting in the middle of the skein. Thus, begin by winding half of the skein into a ball, and start knitting from the remainder of the skein.

Provisionally cast on 68 stitches in scrap yarn; put the distinctive stitch marker after stitch 17, and the others after stitches 34, 51, and 68.

In order to do magic loop knitting, divide the stitches into two equal groups of 34 stitches each. The first group consists of stitches 18–51 and the second group consists of stitches 52–68 and 1–17. So, pull out some of the excess cable from your needles between stitches 17 and 18, and between stitches 51 and 52. Now begin magic loop knitting!

Purl stitches 1–34 and knit stitches 35–68. Now begin following the charts (from right to left); a key is given in Figure 4. Your distinctive stitch marker marks the beginning of a new row of Figure 6. Notice that each chart is for half of the stitches; so, work one row of the Figure 5 chart, and then one row of the Figure 6 chart before progressing to the next row of the Figure 5 chart. Also notice that the first row of the Figure 5 chart begins in the middle; you will purl 17 stitches before switching to the chart of Figure 6.

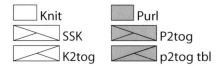

Figure 4. The key for Figures 5 and 6.

When the charts are complete, there will be eight stitches left on the needle. Cut the yarn (leaving about eight inches), thread the tail through the remaining stitches, pull tight, and hide the end of the yarn.

The self-intersecting portion. To knit the second half of the cross-cap, we need to reorganize the stitches. By changing the order in which they are knit, we switch from knitting a tube to knitting a self-intersecting tube.

Remove the provisional cast-on and place the free stitches on the needle, as follows. Cut the scrap yarn between stitches 34 and 35. Now remove the scrap yarn from stitches 35–68, and put the resulting live stitches onto your active needle tip. Go back to stitch 34, and remove the scrap yarn *in reverse order* from stitch 34 to stitch 1, again putting the live stitches onto the active needle tip. See Figure 7 for an illustration of this arrangement of stitches.

Pull on both needle tips to make stitches 68 and 34 adjacent, and then group your stitches for magic loop knitting: pull out some excess cable between stitches 17 and 18, and between stitches 51 and 52. See Figure 8 for how things should look now.

Renumber the stitches sequentially, in the order you will knit them. (The next stitch you knit will be stitch 1.) Put your distinctive stitch marker after the 17th stitch, and the other markers after stitches 34, 51, and 68. (You may want to mark this row with a locking stitch marker or safety pin on the 51st stitch, to help with counting rows.) Now start purling. From here on, only purl, and never knit—the effect that half is in stockinette stitch and half is in reverse stockinette comes from the self-intersection, not from changing the type of stitch.

Near the point of intersection, the passive needle tip (holding old stitches) should be under the cable of the circular needle. At the point of intersection, switch so that the active needle tip (holding new stitches) now crosses over the cable (see Figure 9). Don't mistake yarn belonging to the intersecting fabric for stitches, or you will start increasing like crazy.

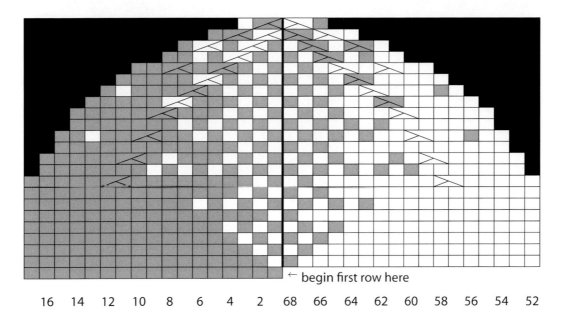

16 14 12 10 8 6 4 2 68 66 64 62 60 58 56 54 52

Figure 5. Chart for stitches 52–68 and stitches 1–17. The vertical black line corresponds to the stitch marker indicating the end/beginning of each round.

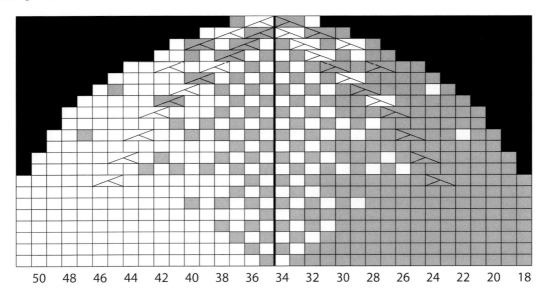

50 48 46 44 42 40 38 36 34 32 30 28 26 24 22 20 18

Figure 6. Chart for stitches 18–51. The vertical black line corresponds to the stitch marker at stitch 34.

Figure 7. Stitches, picked up in order 35–68, then 34 down to 1.

For those who prefer charts over line-by-line instructions, note that the pattern of decreases here is the same as the decreases on the charts for the non-self-intersecting disk. Just ignore the directions to knit some and purl some, and purl every stitch.

Row 1: P17, turn to work on other half of stitches, p17, cross needle cable, p17.

Row 2–7: P17, cross needle cable, p17, turn to work on other half of stitches. Repeat.

Row 8: P5, p2tog, p10, cross needle cable, p10, p2tog tbl, p5, turn to work on other half of stitches. Repeat.

Row 9: P16, cross needle cable, p16, turn to work on other half of stitches. Repeat.

Row 10: P5, p2tog, p9, cross needle cable, p9, p2tog tbl, p5, turn to work on other half of stitches. Repeat.

Row 11: P15, cross needle cable, p15, turn to work on other half of stitches. Repeat.

Row 12: P5, p2tog, p8, cross needle cable, p8, p2tog tbl, p5, turn to work on other half of stitches. Repeat.

Row 13: P14, cross needle cable, p14, turn to work on other half of stitches. Repeat.

Figure 8. Tighten the needles, and pull excess cable out between stitches 17–18 and 51–52 to form a magic figure-8 loop.

Figure 9. Near the intersection, the passive needle tip should be going under the cable. At the intersection, switch so that the active needle tip is now above the cable.

Row 14: P, p2tog, p7, cross needle cable, p7, p2tog tbl, p5, turn to work on other half of stitches. Repeat.

Now is a good time to stuff the cross-cap most of the way.

Row 15: P5, p2tog, p6, cross needle cable, p6, p2tog tbl, p5, turn to work on other half of stitches. Repeat.

Row 16: P5, p2tog, p5, cross needle cable, p5, p2tog tbl, p5, turn to work on other half of stitches. Repeat.

Row 17: P5, p2tog, p4, cross needle cable, p4, p2tog tbl, p5, turn to work on other half of stitches. Repeat.

Row 18: P2, p2tog, p1, p2tog, p3, cross needle cable, p3, p2tog tbl, p1 p2tog tbl, p2, turn to work on other half of stitches. Repeat.

Row 19: P4, p2tog, p2, cross needle cable, p2, p2tog tbl, p4, turn to work on other half of stitches. Repeat.

Top up that stuffing a bit.

Row 20: P1, p2tog, p1, p2tog, p1, cross needle cable, p1, p2tog tbl, p1, p2tog tbl, p1, turn to work on other half of stitches. Repeat.

Row 21: P1, p2tog, p2tog, cross needle cable, p2tog tbl, p2tog tbl, p1, turn to work on other half of stitches. Repeat.

Row 22: P1, p2tog, cross needle cable, p2tog tbl, p1, turn to work on other half of stitches. Repeat.

Eight stitches should remain. Cut the yarn, leaving a foot of tail, and thread it through the remaining stitches. Top up the stuffing one last time, pull tight, and hide the end. Voila!

Bibliography

[1] Anonymous. "How to Knit Small Circumferences Using One Long Circular Needle." Available at http://www.az.com/~andrade/knit/mloop.html, accessed 2009.

[2] Galley, Sara. "Lean on Me." *let me explaiKnit.* Available at http://explaiknit.typepad.com/let_me_explaiknit/2006/04/lean_on_me.html, April 7, 2006.

[3] Vogue Knitting Magazine Editors. *Vogue Knitting: The Ultimate Knitting Book.* Sixth&Spring, New York, 2002.

CHAPTER 4

fashioning fine fractals from fiber

TED ASHTON

1   Overview

G. H. Hardy noted:

> The mathematician's patterns, like the painter's or the poet's, must be *beautiful*; the ideas, like the colors or the words, must fit together in a harmonious way. Beauty is the first test: there is no permanent place in the world for ugly mathematics. [6]

So much of the beauty of mathematics is hidden to the untrained eye. One of the best aspects of fractals is that their beauty is visible even to those who do not understand the math behind them. The Sierpiński triangle (or gasket) not only has that beauty, but the added joy of being simple to create: take any triangle and remove the triangle formed by joining the midpoints of its sides. Now do that to the three smaller triangles that remain. Do it again, this time to the nine triangles just formed. Repeat the process forever. (This method of defining the Sierpiński triangle can be found many places, including [3, pp. 180–181] and [9, p. 12]). Figure 1 illustrates the resulting fractal.

Figure 1. The Sierpiński triangle.

The five fiber fractals in this chapter have as ancestors the crocheted Sierpiński triangles of D. Jacob Wildstrom [16]. His work is based on cellular automata such as Stephen Wolfram describes in *A New Kind of Science* [17]. Applying the same concept to tatting a Sierpiński triangle representation should be simple. However, it

turns out that cellular automata do not fit well with the nature of tatting. Tatted work is constructed primarily of circles and arcs, while cellular automata live in a very square world. This motivated an investigation of mathematical Sierpiński triangle construction methods and subsequently inspired the realization of a natural correspondence between five of these methods and five different fiber arts. This chapter details that correspondence.

In Section 2.1, we link the art of cutwork with the most common (triangle-removing) construction of the Sierpiński triangle, and rue our cutwork naïveté. We then describe properties of fractals in Section 2.2. That leads us, in Section 2.3, to explore a construction that builds the Sierpiński triangle from smaller shapes—*any* smaller shapes. In turn, we use this mathematics to tat and bead Sierpiński triangle representations. In Section 2.4, we construct the Sierpiński triangle by bending a line (ideal for string art). Finally, in Section 2.5, we link cross-stitch to cellular automata.

In Section 3 we use games and artwork to give a hands-on investigation of the concepts introduced in Section 2, for students as young as those in elementary school. The tasks range from creating wearable art to implementing strategies learned from mathematics to succeed at an online game. Some of the activities are suitable for art classwork. Finally, Section 4 gives instructions for tatting and beading the Sierpiński triangle, and for making it with string art and cross-stitch.

2   Mathematics

2.1   Constructing the Sierpiński Triangle by Cutting

Let's build a Sierpiński triangle step by step. For this example we will use a right triangle (as in Figure 2) instead of an equilateral one, though any other triangle would work just as well.

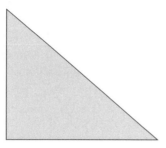

Figure 2. A triangle—one possible beginning for constructing the Sierpiński triangle.

If we join the midpoints of the three sides and remove the open set these line segments bound, we are left with three (closed) triangles, each similar to the original triangle (the side lengths are, of course, each half the length of the corresponding side in the original). Figure 3 shows the result.

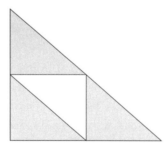

Figure 3. The first "middle" triangle has been removed.

Having had success at our first venture, we do the same thing to each of those three smaller triangles, resulting in nine even smaller triangles (Figure 4). These triangles are, in area, each $\frac{1}{16}$ the size of the original triangle, or $\frac{1}{4}$ the size of each triangle from which they were cut.

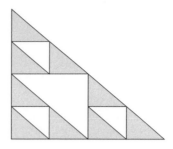

Figure 4. The second step removes the middles of the three triangles created in the first step.

Every time we repeat the process, we find ourselves with three times the number of triangles we had in the previous step, each having half the side length and hence one quarter the area of its immediate ancestor. Figure 5 shows two more steps in the process.

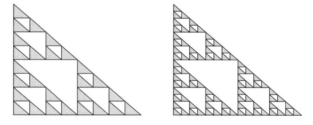

Figure 5. Two more steps in the process.

At each step in the process, the area of the object is $\frac{3}{4}$ the area of the previous iteration. Thus, in the nth step we have an object with area $(\frac{3}{4})^n$ of the original area. As $n \to \infty$, this tends toward zero. In the limit, we find ourselves with the beautiful, lacy object represented in Figure 6.

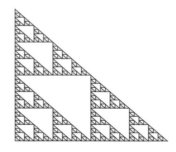

Figure 6. Another Sierpiński triangle (compare to Figure 1).

This method of creating the Sierpiński triangle suggests making a model using cutwork. Sadly, no instructions for creating a cutwork Sierpiński triangle are included. Attempts to create one using white embroidery floss and fabric resulted in artwork (shown in Figure 7) that had to be placed under glass to keep it from falling apart. Should someone create a cutwork representation with structural integrity, please contact the author and/or editors with instructions. (Should a cutwork expert know that this model is impossible, please instead send the explanation for why it cannot be made.)

Figure 7. A cutwork Sierpiński triangle.

2.2 Fractals, Dimension, and a Map

Rather than rigorously defining the term *fractal*, we follow Kenneth Falconer's advice to avoid rigor [5, p. xx]. Instead, we opt to list widely recognized attributes of fractals. The first is self-similarity. Examine a small portion of a fractal and you find a figure that is somehow like the entire fractal (see [9] for lovely illustrations of this or [5] for a mathematical exposition). The next attribute is invariance under a specific transformation. In what follows, we will discover a particular transformation of the plane that turns any set of points into the Sierpiński triangle. Finally, fractals appear to have dimension that is not integral. That will take a little explaining to understand, so before we go into it, let's see how the Sierpiński triangle has the first two attributes.

Double your pleasure, triple your fun. Suppose we were to take a $2\times$ magnifying glass and look at the upper three triangles in Figure 4. We would discover that they form exactly the triangle of Figure 3. The same remains true of later steps—applying $2\times$ magnification to the "upper third" (or either of the other two "thirds") of *any* given step gives us the figure of the previous step. If we can go back one step by doubling the size of one third of the figure, then we can go forward one step by shrinking the entire figure to half its size, making two more copies of the shrunken figure, and arranging the three copies so that they sit at the three corners of the original shape (as shown in Figure 8).

To say this somewhat more mathematically, each of the three constituent triangles is the result of applying a linear map that has the associated vertex as a fixed point and that contracts all distances to one-half their original magnitudes. We'll use the symbol $T_{\cup 3}$ to denote the union of those three contraction mappings.

Notice that the completed Sierpiński triangle, denoted $\mathbb{T}$, is the result of a limiting process, and is distinct from any individual step in the process. When we apply $T_{\cup 3}$ to one of the steps, we get the next step, and each step is different from its predecessor. In particular, it has pieces missing that were there in the previous step. But $\mathbb{T}$ has no more pieces to be removed, and when we apply $T_{\cup 3}$ to $\mathbb{T}$, it doesn't change a bit.

The Sierpiński triangle fits the definition of a fractal as a set that is unchanged by the transformation $T_{\cup 3}$. It is also self-similar: every one of the smaller triangles that makes up a Sierpiński triangle is itself a Sierpiński triangle.

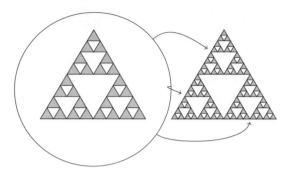

Figure 8. We can move from one step of the construction to the next by making three half-sized copies of the current figure and arranging them to form the new figure.

Dimensions. In his definition of a fractal, Mandelbrot invoked two different ideas of dimension. The first type of dimension is topological dimension. It is always an integer and measures dimension as one tends to think of it—if a figure contains any continuous curves its topological dimension is at least 1. If it contains a filled disk, no matter how small, it will have topological dimension at least 2, and so on.

It may seem surprising at first to know that the Sierpiński triangle has topological dimension 1 (the interested reader can find a proof of this at [4, p. 97]). However, it is less surprising when we notice that every little triangle within $\mathbb{T}$, no matter how small, has its middle removed at some step, so the Sierpiński triangle has no filled disks. As we noted on page 60, the area of the figure vanishes as the step number increases.

Mandelbrot calls the second type of dimension *Hausdorff-Besicovitch* dimension. Consider an idealized spring, which is a spiral in three-dimensional space. It is not a three-dimensional object, but is instead one-dimensional. To see this, start by covering the spring with tiny cubes of side length δ. The number of cubes required depends on δ and so we call it c_δ. Multiply δ by c_δ, and take the limit of $c_\delta \cdot \delta$ as δ goes to 0. For the idealized spring, $c_\delta \cdot \delta$ will remain constant. If the spring is ℓ units long, we can always cover it with at

most $\frac{\ell}{\delta} = c_\delta$ cubes, so no matter how small δ gets, $c_\delta \cdot \delta = \ell$.

However, if we were trying to cover a square of side length ℓ instead of a spring, we would need $c_\delta = \left(\frac{\ell}{\delta}\right)^2$ of the cubes, and as δ got small, the product $c_\delta \cdot \delta = \frac{\ell^2}{\delta}$ would get unboundedly large. In contrast, the product $c_\delta \cdot \delta^2 = \left(\frac{\ell}{\delta}\right)^2 \delta^2 = \ell^2$ stays constant for the square and $c_\delta \cdot \delta^2 = \ell\delta$ goes to zero for the spring. Thus, the Hausdorff-Besicovitch dimension and the topological dimension of the square are both two. Similarly, $c_\delta = \left(\frac{\ell}{\delta}\right)^3$ boxes will be needed to fill a cube of side length ℓ, so $c_\delta \cdot \delta^3 = \left(\frac{\ell}{\delta}\right)^3 \delta^3 = \ell^3$ is constant, and 3 is the correct power of δ; using δ^2 or δ^4 would produce infinity or zero, respectively, when we take the limit as δ approaches zero. In other words, the Hausdorff-Besicovitch dimension is the value k such that the product $c_\delta \cdot \delta^k =$ (number of boxes of side length δ needed to cover the object) $\cdot \delta^k$ neither goes to zero nor gets unboundedly large as δ gets small. It is the power of δ in this expression that determines the Hausdorff-Besicovitch dimension.

Looking back at Figures 3–6, we can see that if we let $\delta = \frac{1}{2}$ then $c_\delta = 3$ and if $\delta = \frac{1}{4}$ then $c_\delta = 9$. More generally, in the nth iteration if we let $\delta = \left(\frac{1}{2}\right)^n$, then $c_\delta = 3^n$ because each square of side length $\left(\frac{1}{2}\right)^n$ covers one of the 3^n triangles. We wish to find the value of k that will

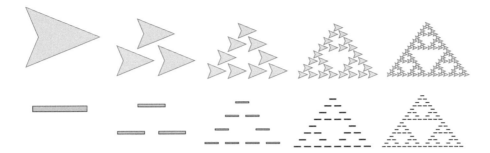

Figure 9. When we take an arbitrary shape and repeatedly apply the triple-contraction map $T_{\cup 3}$ to it, the overall figure seems more and more like a Sierpiński triangle.

make the $\lim_{\delta \to 0} c_\delta \cdot \delta^k$ a nonzero constant. Notice that as δ approaches zero, n approaches infinity. Therefore, we determine the value of k that makes $\lim_{n \to \infty} \left(\frac{3}{2^k}\right)^n$ a nonzero constant. Namely,

$$\frac{3}{2^k} = 1 \implies 3 = 2^k \implies k = \log_2 3 \approx 1.585.$$

This means that the Sierpiński triangle has Hausdorff-Besicovitch dimension $\log_2 3$.

2.3 More Precise Definitions of $T_{\cup 3}$ and $\mathbb{T}$

What if we start with an arbitrary shape and repeatedly apply $T_{\cup 3}$ to it? Figure 9 indicates that the result looks like the Sierpiński triangle.

However, to show that the resulting figure *is* a Sierpiński triangle, we will need a more precise definition of $\mathbb{T}$.

Following a suggestion by Manfred Schroeder [11, p. 22], we give a triangle, $\triangle$, three coordinates: u, v, and w. Each of these coordinates is associated with one of the vertices of the triangle. It takes the value 1 at its vertex and 0 at the opposite side. Each coordinate changes linearly, and its level curves are lines parallel to the side on which it is 0 (see Figure 10).

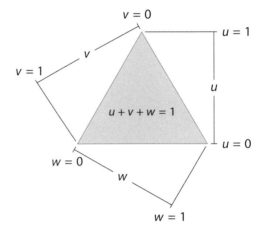

Figure 10. The coordinate system shown on an equilateral triangle (the coordinate system itself is independent of the shape or size of the triangle).

We have three coordinates on a two-dimensional object, so they must be mathematically related. Indeed, $u + v + w = 1$ at any point in $\triangle$. Suppose we look at the edge where $v = 0$ and u runs from 0 to 1 while w runs from 1 to 0. It follows that $w = 1-u$. Similarly, on the side where $w = 0$, $v = 1 - u$. Now for a fixed $u_0 \in [0, 1]$, the line segment that lies within $\triangle$ is the segment running from $(u_0, 1 - u_0, 0)$ to $(u_0, 0, 1 - u_0)$. Because the coordinate change is linear, $w = 1 - u_0 - v$ or $u_0 + v + w = 1$. This is true for all $u \in [0, 1]$, so $u + v + w = 1$. The triangle $\triangle$ is simply the points for which $0 \leqslant u, v, w \leqslant 1$.

Now that we have coordinates on $\triangle$, we can express $T_{\cup 3}$ in terms of those coordinates:

$$T_{\cup 3}(u, v, w) = \left(\frac{u}{2} + \frac{1}{2}, \frac{v}{2}, \frac{w}{2}\right) \cup \left(\frac{u}{2}, \frac{v}{2} + \frac{1}{2}, \frac{w}{2}\right)$$

$$\cup \left(\frac{u}{2}, \frac{v}{2}, \frac{w}{2} + \frac{1}{2}\right)$$

(for a picture of the effect of $T_{\cup 3}$, see Figure 11).

Figure 11. $T_{\cup 3}$ maps one point to three. Here we show $T_{\cup 3}(\frac{3}{4}, \frac{1}{8}, \frac{1}{8}) = (\frac{7}{8}, \frac{1}{16}, \frac{1}{16}) \cup (\frac{3}{8}, \frac{9}{16}, \frac{1}{16}) \cup (\frac{3}{8}, \frac{1}{16}, \frac{9}{16})$.

In looking at $\triangle$ under these coordinates, four regions are naturally created. First, there are the points of the triangle nearest $(1, 0, 0)$, where $u \geqslant \frac{1}{2}$ while $v < \frac{1}{2}$ and $w < \frac{1}{2}$. Similarly, the points of the triangle nearest $(0, 1, 0)$ are those with $v \geqslant \frac{1}{2}$ and the points of the triangle nearest $(0, 0, 1)$ have $w \geqslant \frac{1}{2}$. Finally, in the central subtriangle, u, v, and w are all strictly less than $\frac{1}{2}$. All four regions are shown in Figure 12.

We now look at the three smaller triangles that must be removed in the second step. As we noted at the beginning of Section 2.2, the triangles removed by step 2 are the images under $T_{\cup 3}$ of the triangle removed in step 1, that is $\{\frac{1}{2} < u < \frac{3}{4}; 0 < v, w < \frac{1}{4}\} \cup \{\frac{1}{2} < v < \frac{3}{4}; 0 < u, w < \frac{1}{4}\} \cup \{\frac{1}{2} < w < \frac{3}{4}; 0 < u, v < \frac{1}{4}\}$.

This description will become progressively more complicated at each iteration. Using binary notation is useful in this context, because numbers greater than or equal to $\frac{1}{2}$ become visible as those with a 1 in the first place after the binary point, and numbers less than $\frac{1}{2}$ become visible as those with a 0 in that position.

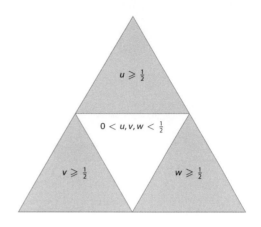

Figure 12. The four regions of $\triangle$.

Consider the first step in the process of creating $\mathbb{T}$. The points that remain are those points that have a 1 in the $\frac{1}{2}$s place in *exactly one* coordinate, such as $(0.1, 0.\overline{01}, 0.00\overline{10})$ or $(0, 0.01, 0.10\overline{1})$. (Recall that a bar over a number or set of numbers means that those numbers repeat ad infinitum.) In binary $1 = 0.11111\ldots$, so the vertices of $\triangle$ are $(0.\overline{1}, 0, 0)$, $(0, 0.\overline{1}, 0)$, and $(0, 0, 0.\overline{1})$. Similarly, we can express the midpoints of the unit equilateral triangle as $(0.1, 0.0\overline{1}, 0)$, $(0.1, 0, 0.0\overline{1})$, and $(0, 0.1, 0.0\overline{1})$. The regions from step 2 are points where u, v, and w all have 0 in the *second* binary place. More generally, $T_{\cup 3}$ first divides every coordinate by 2, which shifts every binary digit one place to the right (leaving zeroes in the first binary place in all coordinates). Then $T_{\cup 3}$ trifurcates, adding $\frac{1}{2}$ to a different coordinate of each copy. That simply changes the 0 in the first binary place of the given coordinate to a 1 (see Figure 13).

Beading the Chaos Game. The Chaos Game [1] uses $T_{\cup 3}$ but at each step we only follow one of the three branches. To set up the game, select three (non-collinear) points in the plane as the vertices of the Sierpiński triangle-to-be and choose a point on the boundary of the triangle these points define. (For convenience, choose all points to have rational coordinates.)

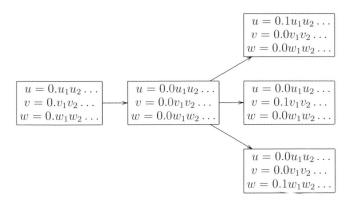

$$u = 0.1u_1u_2\ldots$$
$$v = 0.0v_1v_2\ldots$$
$$w = 0.0w_1w_2\ldots$$

$$u = 0.u_1u_2\ldots$$
$$v = 0.v_1v_2\ldots$$
$$w = 0.w_1w_2\ldots$$

$$u = 0.0u_1u_2\ldots$$
$$v = 0.0v_1v_2\ldots$$
$$w = 0.0w_1w_2\ldots$$

$$u = 0.0u_1u_2\ldots$$
$$v = 0.1v_1v_2\ldots$$
$$w = 0.0w_1w_2\ldots$$

$$u = 0.0u_1u_2\ldots$$
$$v = 0.0v_1v_2\ldots$$
$$w = 0.1w_1w_2\ldots$$

Figure 13. The compound action of $T_{\cup 3}$ on a point is to shift the coordinates and then trifurcate.

To play the game, randomly choose a vertex of the triangle, move halfway from the current position to the chosen vertex, and mark that spot as the new current position. This choice of vertex determines the branch of $T_{\cup 3}$ we follow. Repeat ad infinitum. Every new point produced by this method will be distinct from each that went before it, since on every step the binary representations of u (and v) will be one place longer. This process generates the Sierpiński triangle with probability one [11]. Figure 14 shows two triangles created using the Chaos Game. Instructions for a beaded version of the Chaos Game are in Section 4.2.

Defining $\mathbb{T}$. In the first step of creating the Sierpiński triangle, the removed triangle is composed of points where u, v, and w are zero in the first binary place. That's neat, because now we can say that after the first step in creating $\mathbb{T}$ (when we have three subtriangles), the points remaining each satisfy $u+v+w = 1$ and each has a 1 in the $\frac{1}{2}$s place in exactly one coordinate. Applying $T_{\cup 3}$ to the removed points yields new points whose coordinates are zero in the second binary place, and these are the points removed in the second iteration of making the Sierpiński triangle. In turn, applying $T_{\cup 3}$ n times to the points of the original removed triangle yields the triangles that are removed in the nth iteration—and these points have coordinates that are zero in the nth binary piece of each coordinate. This leads to

Definition 1 The *Sierpiński triangle* $\mathbb{T}$ is those points (u, v, w) such that $0 \leqslant u, v, w \leqslant 1$, $u + v + w = 1$, and in no binary position do u, v, and w all have zeros.

Figure 14. Two triangles created by the Chaos Game. Each has 1,000 points.

About Fractions in Binary

In binary, instead of using the digits 0–9, we only use 0 and 1. The rightmost position of an integer in binary notation is the 1s place, just as with decimal, but the position just to its left expresses how many 2s there are in the number, not how many 10s it has. Similarly, the next position tells how many 4s there are (because $4 = 2^2$) instead of 100s and the next one tells how many 8s there are, and so on.

Having expressed integers in binary notation, we can also express nonintegers in binary. Just as is done in decimal notation, we place a period, called the *binary point*, to the right of the 1s place. The first position to the right of the binary point is the $\frac{1}{2}$s place. The next one is the $\frac{1}{4}$s place, and so on. As an example, the binary number 100.001 is the number $4\frac{1}{8}$ or 4.125 in decimal.

In binary notation, multiplication by a power of 2 or by a power of $\frac{1}{2}$ shifts the binary point to the right or to the left some number of places, just as multiplication by powers of 10 and $\frac{1}{10}$ does in decimal.

With our new understanding of $T_{\cup 3}$ and our new definition of $\mathbb{T}$, we can show that a Sierpiński triangle may be constructed from any shape via repeated application of the $T_{\cup 3}$ map. Let $S \neq \emptyset$ be our original shape and consider $T_{\cup 3}^n(S) = T_{\cup 3}(T_{\cup 3}(\cdots T_{\cup 3}(S)))$. For convenience, assume that $S \subset \triangle$.

Theorem 1 *The shape $T_{\cup 3}^\infty(S) = \lim_{n \to \infty} T_{\cup 3}^n(S)$ is the Sierpiński triangle.*

Proof

Claim 1: $T_{\cup 3}^\infty(S) \subseteq \mathbb{T}$.

Consider $s \in S$. Because $S \subset \triangle$, $s = (u_s, v_s, w_s)$ with $0 \leq u_s, v_s, w_s \leq 1$ and $u_s + v_s + w_s = 1$. These properties also hold for $T_{\cup 3}^i(s)$ and therefore for $T_{\cup 3}^\infty(s)$. We need only show that no point of $T_{\cup 3}^\infty(s)$ has all zeroes in the same binary position of all three coordinates. Suppose for the sake of contradiction that some point of $T_{\cup 3}^\infty(s)$ does have all zeroes in the nth binary position, and examine any $x \in T_{\cup 3}^\infty(s)$ where n is minimal. If we apply $T_{\cup 3}$ to $T_{\cup 3}^\infty(s)$, we see that $T_{\cup 3}(x)$ no longer has all zeroes in the nth binary position. In fact, $T_{\cup 3}(T_{\cup 3}^\infty(s)) = T_{\cup 3}^\infty(s)$, so if $x \in T_{\cup 3}(T_{\cup 3}^\infty(s))$ has all zeroes in the nth binary position, then $x \in T_{\cup 3}^\infty(s)$ had all zeroes in the $(n-1)^{\text{st}}$ binary posi-

tion. This contradicts the minimality of n, and therefore no such point can exist.

Claim 2: $\mathbb{T} \subseteq T_{\cup 3}^\infty(S)$.

Consider again $s \in S$. Notice that $T_{\cup 3}(s)$ is a set of three points with all allowable combinations of 1s and 0s in the first binary place. Similarly, $T_{\cup 3}(T_{\cup 3}(s))$ is a set of nine points with all allowable combinations of 1s and 0s in the first and second places. In general, $T_{\cup 3}^n(s)$ is a set of 3^n points with all allowable combinations of 1s and 0s in the first n places. So, given any point $t \in \mathbb{T}$, there will be a point of $T_{\cup 3}^n(s)$ that matches the coordinates of t in at least n places. This is true for any arbitrarily large n, so in the limit there exists a point of $T_{\cup 3}^\infty(S)$ that is equal to t. □

Tatting $T_{\cup 3}^n(S)$. The triple contraction map, $T_{\cup 3}$, is the mathematical technique needed to tat the Sierpiński triangle. We begin by tatting a small triangle (made of circles) and assemble three copies of this small triangle into a larger triangle. (Instructions are in Section 4.1.) This is effectively an application of $T_{\cup 3}$ to the small triangle. Of course, once tatted, a triangle can't be contracted to half its size, but human perception takes care

Figure 15. As the tatted triangle grows by the tripling map, each circle becomes a smaller part of the whole.

of that problem for us. As the entire structure grows in size, each individual triangle becomes a smaller part of the whole (see Figure 15).

2.4 Bending a String

Recall Definition 1 of $\mathbb{T}$ and fix u with terminating binary representation of length ℓ (e.g., $u = 0.1011$ where $\ell = 4$). Now, line segments contained in the Sierpiński triangle can be constructed as follows: fix a value for the first ℓ places of v, requiring that where u has a 1, v must have a 0. Using that fixed value for the first ℓ places of v, let the remaining places vary over all possible values. Then, $(u, v, 1 - u - v)$ will trace out a line segment of length $\frac{1}{2^\ell}$.

Of course, the same holds true if v or w is the coordinate with the terminating binary representation. Therefore, much of the Sierpiński triangle lies on line segments. In fact, a point in which none of the coordinates has a terminating representation can be approximated arbitrarily closely by a point in which one of the coordinates does have a terminating representation. (This process is analogous to the approximation of reals by rationals.)

A sequence of such line segments could be used to approximate the Sierpiński triangle with a single curve. It appears most natural to use the outlines of triangles as our line segments. Building $\mathbb{T}$ by outlining stages of construction using longer and longer curves may seem straightforward (see Figure 16). However, a parameterization for one level of the curve need not extend to a parameterization of the next level and so this curve is mathematically unnatural. There is a more natural mathematical method for creating the Sierpiński triangle from a single parameterized curve [2], but it turns out to be impractical to realize as string art. The pattern given in Section 4.3 creates an approximation to $\mathbb{T}$ using the curves given in Figure 16 (but without computing a parameterization).

Notice that the Sierpiński triangle can be constructed either by starting with a two-dimensional object and removing pieces, or by starting with a one-

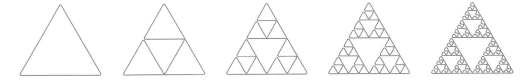

Figure 16. Creating the triangle by outlining more and more detail. All corners have been bent slightly to emphasize the path.

dimensional object and bending it. This gives new meaning to the term "fractional dimension."

2.5 Cross-Stitching Cellular Automata

A cellular automaton iterates a collection of rules to create an evolving picture of a grid of boxes. Here, we will be concerned with a rule that acts on an infinite row of boxes. Each box has one of two states (empty or filled). Every box has a neighborhood, consisting of the box itself, the box to its left, and the box to its right (see Figure 17).

Neighborhood

Figure 17. Possible input to a cellular automata. One neighborhood is marked.

Each box can be either filled or empty, so any given neighborhood can be in one of eight states, running from all-empty to all-full. The neighborhood of a box in one line of a grid determines the state of the corresponding box in the next line of the grid, where grid lines correspond to successive states of the automaton. As shown in Figure 18, this creates a rule determining the cellular automaton; the input states are considered as binary numbers with the filled boxes acting as 1s

and the empty boxes acting as 0s. As there are eight different neighborhood states and each can result in a filled box or an empty one, there are 256 rules; they are uniquely identified by summing the values of the labels of full boxes. The rule shown in Figure 18 is rule $2^1 + 2^4 = 2 + 16 = 18$.

For example, suppose we make the choices shown in Figure 18 and start with a row that has three consecutive filled boxes—all the others are empty.

Considering the neighborhoods from left to right in succession, the top row of Figure 19 has several 2^0 neighborhoods followed by a 2^1 neighborhood, a 2^3 neighborhood, a 2^7 neighborhood, a 2^6 neighborhood, a 2^4 neighborhood, and then more 2^0 neighborhoods. The second row of Figure 19 shows the result of applying Rule 18 to each of the neighborhoods.

If we start with an impulse (a row with a single filled box) and apply Rule 18 repeatedly, we get the pattern shown in Figure 20.

In examining Figure 20, we see that some parts of Rule 18 are used, while others are not. For example, at no point in the pattern does there ever exist a neighborhood in which all of the boxes are full (a 2^7 neighborhood). In fact, no two filled boxes are ever adjacent to each other, so the 2^7, 2^6, and 2^3 portions of the rule are never used. Hence, there are actually eight different rules that would create the same pattern: 18, 26, 82, 90,

Figure 18. One possible rule (Rule 18).

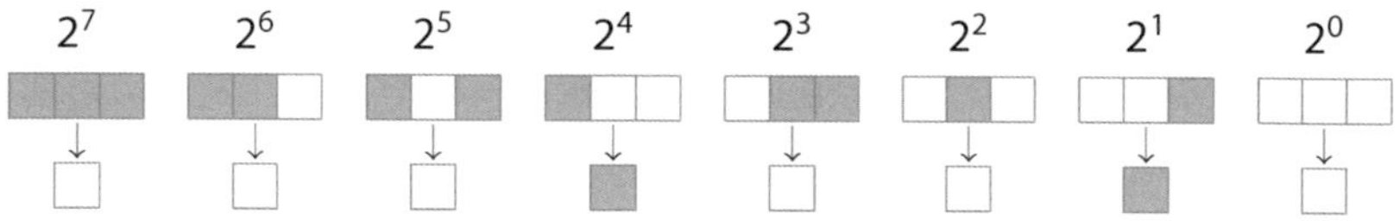

2^7 $\quad$ 2^6 $\quad$ 2^5 $\quad$ 2^4 $\quad$ 2^3 $\quad$ 2^2 $\quad$ 2^1 $\quad$ 2^0

Figure 19. The successive rows show repeated application of Rule 18. Time progresses down the page.

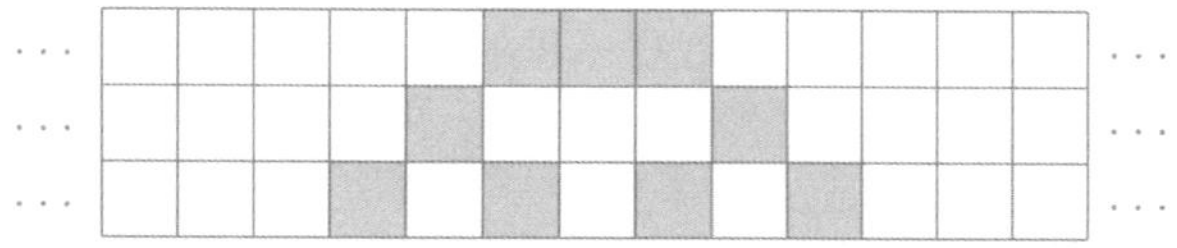

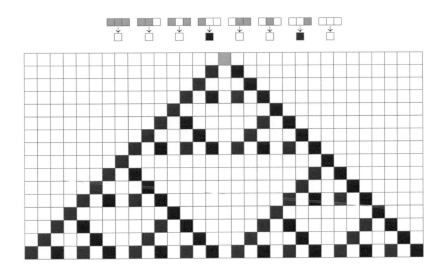

Figure 20. The pattern created by repeatedly applying Rule 18 to an impulse. The filled boxes are colored red or green depending on whether they were created by the 2^1 or the 2^4 portion of the rule.

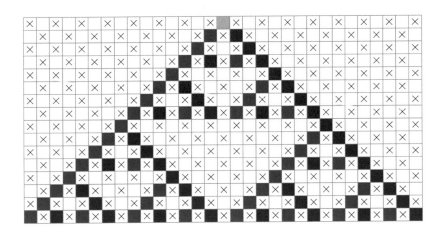

Figure 21. The squares marked with $\times$'s are always empty and so we can consider them to be outside of the ambient space of the figure.

146, 154, 210, and 218. This does not imply the equivalence of these eight rules; some starting inputs will produce different behavior for these rules.

Figure 20 is surprisingly reminiscent of the Sierpiński triangle, yet there are significant differences. Notice that $\mathbb{T}$ has a solid lower edge, while half the blocks on the bottom line of Figure 20 are empty, and in fact every other diagonal in the whole picture is entirely empty (see Figure 21). In Figure 22 we remove them completely.

Figure 22. The figure created by Rule 18 *in its ambient space* does have solid horizontal lines.

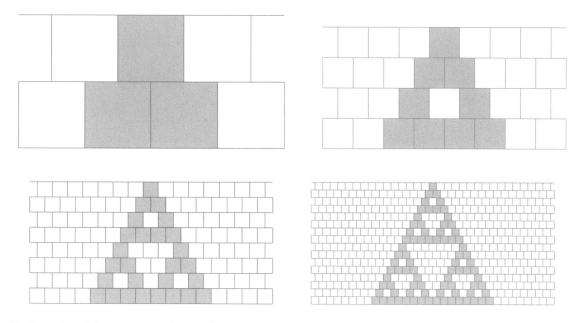

Figure 23. The only cellular automaton figures that are candidates for being Sierpiński triangles are those with 2^n rows. Here they are shown on increasingly fine grids to echo the other progressions we have seen.

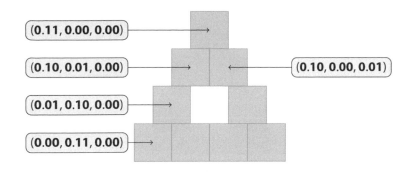

Figure 24. The coordinates of some of the nine blocks in the four-row triangle.

Of course, even with the newly solid lines, the figure still may or may not represent a Sierpiński triangle. To decide whether it does, we need to establish (u, v, w) coordinates on it and see if our definition fits. Clearly, if the figure has some number of rows that is not a power of 2, it will be only a partial triangle—the solid horizontal lines only come every 2^n rows. Restricting our attention to grids with 2^n rows, we get images like those in Figure 23.

We first turn our attention to the top-left shape in Figure 23. Here there only are three colored blocks. In our coordinate system, the top block will be $(0.1, 0, 0)$, the lower left one $(0, 0.1, 0)$, and the one on the lower right $(0, 0, 0.1)$. In the next stage, we have nine colored blocks with coordinate values as shown in Figure 24.

In general, the nth stage will have coordinates whose binary representations truncate after n binary places and whose values run from $0.0\cdots0$ to $0.1\cdots1$ with $u + v + w = 0.1\cdots1$. Extending a figure from 2^n rows to 2^{n+1} rows has the same effect on these coordinates as applying $T_{\cup 3}$ to them, which the studious reader is encouraged to verify.

We therefore conclude that these figures do represent the Sierpiński triangle. As $2^n \to \infty$ we have that

$u + v + w \to 0.111\ldots = 1$ and that the figure approaches the Sierpiński triangle in essentially the same way that Figure 15 does.

Figure 25. The Sierpiński triangle, cross-stitched from Rule 18.

While D. Jacob Wildstrom used crochet to make his cellular-automata triangles [16], the grid of Figure 20 also suggests cross-stitch; instructions for several cross-stitched cellular automata Sierpiński triangles are given in Section 4.4.

3 Teaching Ideas

In this section, ideas are given to help students explore in depth the mathematics related to each Sierpiński triangle construction. Section 3.1 encourages analysis of the pointwise (single-branch $T_{\cup 3}$) construction using paper, as well as use of the binary coordinates to reach given points efficiently. In contrast, Section 3.2 uses the $T_{\cup 3}$ construction to morph a given image into the Sierpiński triangle, and thereby highlights the difference between a limiting process and its resulting limit. Section 3.3 suggests variations of our original cellular automaton to more dimensions and different rules.

3.1 The Chaos Game

Students can enjoy the Chaos Game starting in late elementary school. Split students into pairs and give each pair a sheet of paper, a pen, a six-sided die, a copy of the rules of the Chaos Game, and the ruler from page 78. Have each pair draw a large triangle on the paper and label the vertices with the numbers 1–6 so that each vertex is assigned two of the numbers. One student should operate the die and the other should draw dots on the paper. The rules of the Chaos Game are as follows:

1. Start at the top vertex of the triangle.

2. Roll the die.

3. Move halfway from the current position to the vertex indicated by the die roll.

4. Make a dot at this spot. It becomes the new current position.

5. Return to step 2.

It takes a few hundred points for the pattern to become very visible, so it is important to make sure that sufficient time and endurance are available. An electronic exploration of this idea is available at [14]. In this vein, computer science students may be interested in attempting to code simple versions of the Chaos Game.

For an extracurricular experience with lasting impact, decorate sweatshirts. For example, math club members may want to make unique, yet matching, sequined or beaded Chaos Game apparel. (Also, fabric paint dots will produce a quicker Chaos Game, and a paper-cutwork Sierpiński triangle can be used to create a silkscreen template for t-shirts.)

A variant of the Chaos Game is available at [13], where the objective is to move a point from a vertex to a given area of the Sierpiński triangle in the least number of moves. Students can use the binary representation given in Section 2.3 to develop an optimal strategy.

3.2 The Triple-Contraction Map $T_{\cup 3}$

A simple demonstration of $T_{\cup 3}$ can illuminate the formation of fractals and create a fascinating work of art. For this activity, a student will need access to a scanner and image-manipulation software, or to a color photocopier. The student should select a beautiful scrap of fabric or bit of yarn and scan this into the computer. Then, this image can be reduced by half and replicated twice, and the three resulting images can be arranged into a triangle. This process can be repeated on the resulting triangle to form a new iteration of the Sierpiński triangle. After some point, the human eye will not be able to distinguish between one iteration and the next and it will appear that the Sierpiński triangle is complete. Indeed, applying $T_{\cup 3}$ produces changes below the resolution of the screen or copier. This illustrates the difference between applying $T_{\cup 3}$ at one of the steps on the way to producing the Sierpiński triangle and applying it to the Sierpiński triangle itself. Applying it to one of the steps creates a new object that is different than the previous one. Applying it to the Sierpiński triangle has no effect whatsoever. The cognitive difference between the stages of construction of the Sierpiński triangle and the completed Sierpiński triangle is highlighted by the fact that the latter is unchanged by application of the $T_{\cup 3}$ map, whereas the former is not.

Figure 26. Coloring chart for Rule 90.

3.3 Cellular Automata in the Classroom

Cellular automata are accessible to students as early as elementary school. Each student will need two sheets of graph paper, four colored pencils, and a chart as shown in Figure 26.

Begin by having the student color in boxes *A, B, C,* and *D* with her four pencil colors. Holding the graph paper in landscape orientation, the student should color in a single box at the center of the top row of boxes on one sheet and two boxes together at the center of the top row on the other sheet. Then the student can follow the chart to color the boxes in the remaining rows. The results on the first sheet should resemble Figure 20.

Questions that arise naturally from this process include:

* Since two different patterns came from the same rule using different beginning patterns, what will result from other starting patterns?

* What patterns are created by changing the rule? There seem to be many different automata that produce apparent Sierpiński triangles. Eric Weisstein has shown the results of applying each of the 256 rules to an impulse at [15]. Students might experiment with

  – Rule 22, which produces a very similar pattern to that of Rule 18,

  – Rules 60 and 102, which produce Sierpiński triangle representations of very different shapes (these are shown in Figure 44),

  – Rules 90, 94, 122, and 126, each of which, from

a *double impulse*—two adjacent filled blocks—will create a Sierpiński triangle representation with double thickness (the resulting pattern is exhibited in Figure 40),

  – Rule 182, which (again from a single impulse) gives a visually different result, shown in Figure 42. If one were to take a Rule 182 triangle and superimpose a Rule 18 triangle on it, starting one row higher, the two would mesh perfectly to create a solid triangle.

College or graduate students might use Definition 1 to explore how well any or all of the patterns created by the above rules qualify as Sierpiński triangle representations.

3.4 Project Ideas

Here are three open-ended ideas for student investigation.

* There is a relationship between cellular automata and Pascal's triangle, which is explored at [18] and in [16, Section 3]. Challenge: cross-stitch the Sierpinski triangle in colors so as to highlight the math in Pascal's triangle.

* What if a cellular automaton neighborhood were defined to be five squares long instead of three? How many possible rules are there? Do any of them produce an image that looks like a Sierpiński triangle? Do any qualify as actual Sierpiński triangles by Definition 1?

⋆ What if we work with two-dimensional cellular automata instead of one-dimensional cellular automata? The two-dimensional analog of the automaton of Section 2.5 is John Conway's Game of Life, which provides a rich field of study. It can be explored at many sites, such as [7] or [8]. The open-source package Golly [2] can not only run two-dimensional Game-of-Life variants, but can also step through one-dimensional cellular automata. In the Game of Life, using rule 12/1 creates a pattern that appears to be made of four copies of the Sierpiński triangle. Is it? Are there other rules with similar effects?

4 Crafting the Sierpiński Triangle

4.1 A Tatted Sierpiński Triangle

Materials

⋆ One tatting shuttle.

⋆ Artist's choice of tatting thread.

The Basic Unit

The tatted Sierpiński triangle (shown in Figure 27) is created entirely out of one shape, a tiny triangle made of three rings (see Figure 28).

Figure 27. A tatted Sierpiński triangle seen by rubber monsters.

Figure 28. Here is a tiny triangle formed from three rings along with the start of the next tiny triangle.

We start with summary instructions, then follow with illustrations:

Ring 1: Three double stitches, five picots separated by three double stitches, three double stitches, close.

Ring 2: Three double stitches, join to last picot of Ring 1, three double stitches, four picots separated by three double stitches, three double stitches, close.

Ring 3: Three double stitches, join to last picot of Ring 2, three double stitches, three picots separated by three double stitches, three double stitches, join to first picot of Ring 1 as in Figure 29, three double stitches, close and cut thread.

Instructions

The rings should be made with no space between them on the thread and the picots should be short enough to hold the rings together in a tight triangle. The last join,

joining Ring 3 to Ring 1, can be tricky at first. The tiny triangle has nine free picots around it, three on each ring. When joining two tiny triangles together, two of those picots are used for structural integrity. (One might think only one should be used for mathematical integrity.)

The second small triangle is like the first, except for its joins to the first triangle:

Ring 1: Three double stitches, picot, three double stitches, join to the final picot made on the first triangle, three double stitches, join to the next-to-last picot on the first triangle, three double stitches, two picots separated by three double stitches, three double stitches, close.

Rings 2 and 3: Proceed as for the first tiny triangle.

The third tiny triangle differs from the second only in joins on the third ring:

Ring 1: Proceed as for the second tiny triangle.

Ring 2: Proceed as for the first tiny triangle.

Ring 3: Three double stitches, join to last picot of Ring 2, three double stitches, picot, three double stitches, join to third picot made on the first tiny triangle, three double stitches, join to the second picot made on the first tiny triangle, three double stitches, join to first picot of Ring 1, three double stitches, close and cut thread.

All three of those final joins use the same technique as pictured in Figure 29, curving the ring around to bring the ring thread under the picot of interest and making a normal join.

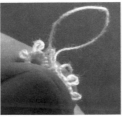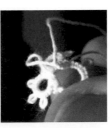

Figure 29. To make the final join, curve the third ring (with its three picots) around and lay the thread under the first picot of the first ring. Pull it up through the picot and pass the shuttle through it to make the join.

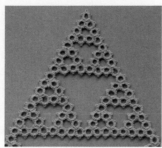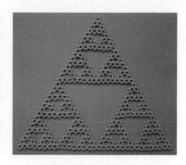

Figure 30. Three different iterations of the same Sierpiński triangle.

At each size, to make a Sierpiński triangle larger, create two more copies of the triangle you already have, joining in the appropriate picots. Examples are shown in Figure 30.

When the desired triangle has been completed, the piece is finished by clipping the knotted ends short or using Lily's Way [10].

Done in a small thread, the tatted Sierpiński triangle could serve as a lace appliquè or inset, or it could make a lovely doily. Presumably, with large enough thread (or enough triangle), it could be made into a lace shawl. At that size, however, the missing center triangle would compromise the strength and shape of the shawl. One might solve this by backing the triangle with cloth of a contrasting color or by filling in the center space using a tatted triangle one quarter the size of the whole, perhaps in another color so as to preserve the mathematical integrity of the tatted fractal.

4.2 A Cloth-and-Bead Sierpiński Triangle

Materials

* A six-sided die.

* A few hundred beads.

* Some plain-colored lightweight cloth.

* A mechanism for stretching the fabric flat, such as an embroidery hoop.

* Needle and thread, or glue, for attaching the beads to the cloth.

* Masking tape (or paper and pins) and pen for labels.

* The ruler given in Figure 33.

* Optional, for making a densely beaded Sierpiński triangle: needle-nose pliers to help pull the needle through the cloth.

The instructions for beading a Sierpiński triangle as in Figure 31 essentially say "attach beads to cloth." The method of doing so, as well as the choice of beads and cloth, is left up to the artist. If sewing the beads to cloth, take a tiny stitch after placing each bead to help secure the thread, and be careful not to make long stitches between beads too tight (or too loose). Note: beginners should use either a large triangle or small beads. A ruler designed to be enlarged for use with larger triangles is provided in Figure 34.

Instructions

Attach three beads to the cloth to form the vertices of the triangle. Any three points will do, as the triangle need not be equilateral.

Figure 31. The Chaos Game applied to beading results in emerging Sierpiński triangles.

For an equilateral triangle, attach two of the beads and note the distance d between them. Now fold the cloth so that the two beads match up. Locate the point that is both on the fold *and* a distance d from the touching beads by placing the end of a measuring tape at the beads and swinging it until the right length hits the fold; place the third bead at this point. These three beads will form an equilateral triangle.

You may wish to outline the triangle with tape as shown in Figure 32 and mark the midpoints of the sides for reference. Label (with pin or tape) one of the vertices "1, 6," one of them "2, 5," and one of them "3, 4." The vertex marked "1, 6" will be your starting point. Place a bead at each vertex.

Roll the die until you get some number other than 1 or 6. Attach a bead to the cloth halfway between the current bead (the "1, 6" bead) and the bead that has the number you rolled (if you rolled a 3, that would be the "3, 4" bead; if you rolled a 5, it would be the "2, 5" bead, and so on).

Roll the die again. Attach a new bead halfway between the bead just attached and the bead with the number just rolled. For added precision, use the Figure 33 ruler: lay the ruler on your work so that the bead you just attached and the bead the die indicates are at matching numbers (e.g., one at each of the lines numbered 12); the center line then indicates the halfway mark.

Repeat this process (roll the die, attach a new bead halfway between where you are and the bead you rolled) until satisfied with the resulting triangle. If sewing the beads to the cloth, consider reducing the amount of running stitch by using three needles, one associated to the area near each vertex.

Figure 32. A finished Sierpiński triangle outlined in tape.

Figure 33. Ruler used to help in locating the halfway point between the most recently placed bead and the next selected vertex.

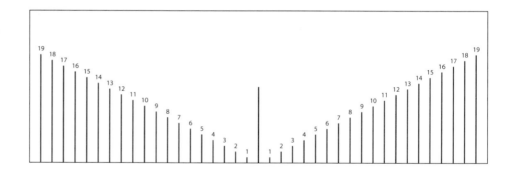

Figure 34. Enlarge this ruler for use with large triangles.

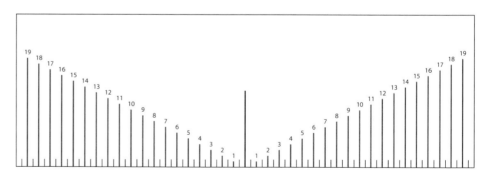

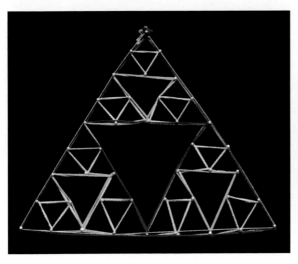
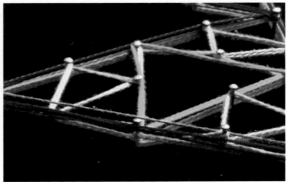

Figure 35. A string art representation of the Sierpiński triangle; note in the detail photo that larger subtriangles have more layers and colors of string.

It will probably take a couple hundred beads before the four largest holes are really evident and 500–1,000 beads before the process is complete. Those weary of rolling the die may wish to have a computer do the job, perhaps by writing a tiny script to create a page full of random 1s, 2s, and 3s that substitute for die rolls.

4.3   The String Art Triangle

Specific material sizes for a string art Sierpiński triangle as in Figure 35 are given here, though the triangle can be made at any size by enlarging or reducing Figure 36.

Materials

* A square of wood, at least $10'' \times 10'' \times 0.5''$. Such a square can be cut from a length of 1×12 or 1×16.

* Half a yard of black velvet.

* Thumbtacks.

* Sewing pins.

* At least 45 brass nails.

* Four colors of embroidery floss (choice of colors left to the artist).

Instructions

Make a copy of Figure 36, enlarging it to about 210% of its size here (8″ wide). Cut the velvet to $18'' \times 18''$ (or 3″ larger per side than your wood square). Cover the front of the board with it, wrapping the edges around the board and using the thumbtacks to attach them to the back. Center the copy of Figure 36 on the velvet (nail 13 is close to the center) and pin it to the fabric. Drive the brass nails through the figure and velvet, partway into the board, and leave at least $\frac{1}{4}''$ of each nail exposed. For aesthetic purposes, each nail head should be the same distance above the velvet. For each color of floss, using two strands at a time, tie onto nail 1, wrap it around the nails in the order given in the following chart, and then tie it off on nail 1. The paper can either be removed prior to adding the floss (in which case, refer to the figure in the book for the nail numbers) or afterward.

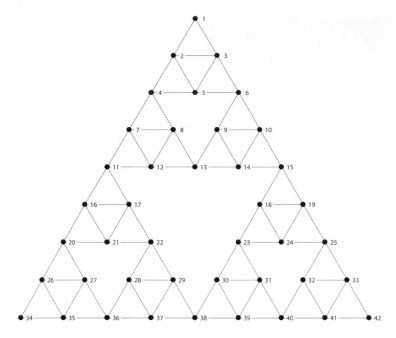

Figure 36. A pattern for the string art Sierpiński triangle.

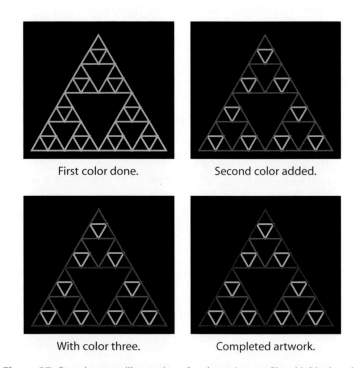

First color done.

Second color added.

With color three.

Completed artwork.

Figure 37. Step-by-step illustrations for the string art Sierpiński triangle.

Color 1: 1 – 42 – 41 – 33 – 32 – 41 – 40 – 25 – 24 – 19 –
18 – 24 – 23 – 40 – 39 – 31 – 30 – 39 – 38 – 15 –
14 – 10 – 9 – 14 – 13 – 6 – 5 – 3 – 2 – 5 – 4 – 13 –
12 – 8 – 7 – 12 – 11 – 38 – 37 – 29 – 28 – 37 –
36 – 22 – 21 – 17 – 16 – 21 – 20 – 36 – 35 – 27 –
26 – 35 – 34 – 1

Color 2: 1 – 42 – 40 – 25 – 23 – 40 – 38 – 15 – 13 – 6 –
4 – 13 – 11 – 38 – 36 – 22 – 20 – 36 – 34 – 1

Color 3: 1 – 42 – 38 – 15 – 11 – 38 – 34 – 1

Color 4: 1 – 42 – 34 – 1

See Figure 37 for how the string art should look after each step.

4.4 Cross-Stitch

Materials

To make any one of the following cross-stitch patterns, some of which are shown finished in Figure 38, you will need:

⋆ 11-, 14-, 16-, or 18-count Aida cloth.

⋆ Embroidery floss. Colors are listed with the individual patterns. All stitches are made using two strands. One skein of each color should prove more than sufficient.

The approximate measurements for the final designs are given in the table below.

Instructions

Basic cross-stitch instructions as well as finishing techniques are given in Chapter 6. Figures 39, 40, 42, 44, and 46 give instructions for different instantiations of a cross-stitched Sierpiński triangle; finished objects are shown in Figures 41, 43, 45, and 47. In each diagram, a different color is used for each part of each rule. For aesthetic reasons, some artists may wish to use the same color for multiple rule-parts/symbols.

Figure 38. Three cross-stitched Sierpiński triangles accompanied by eggs.

Thread count	Measurements for Figure 39, Figure 40 and Figure 42	Measurements for Figure 44
11-count	$6'' \times 3''$	$3'' \times 3''$
14-count	$4\frac{1}{2}'' \times 2\frac{1}{4}''$	$2\frac{1}{4}'' \times 2\frac{1}{4}''$
16-count	$4'' \times 2''$	$2'' \times 2''$
18-count	$3\frac{1}{2}'' \times 1\frac{3}{4}''$	$1\frac{3}{4}'' \times 1\frac{3}{4}''$
stitch count	64×32	32×32

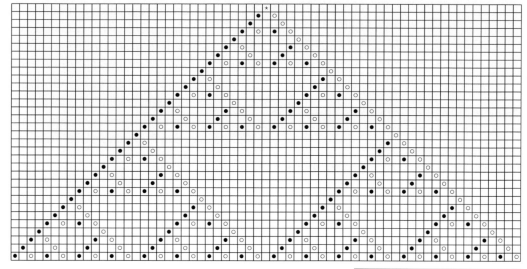

Figure 39. Cross-stitch pattern for triangle created by Rules 18, 26, 82, 90, 146, 154, 210, and 218 from an impulse. A finished object is shown in Figure 25.

★ DMC 783 – Medium Topaz
● DMC 561 – Very Dark Jade
○ DMC 816 – Garnet

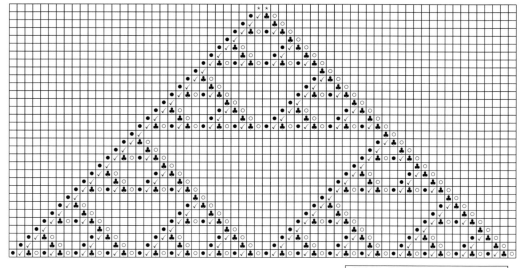

Figure 40. Rules 90, 94, 122, and 126 yield this pattern from a double impulse. If stitched in the listed colors, this produces an optical illusion that causes the finished piece to appear blurry.

★ DMC 318 – Light Steel Gray
● DMC 603 – Cranberry
✓ DMC 498 – Dark Red
♣ DMC 336 – Blue
○ DMC 813 – Light Blue

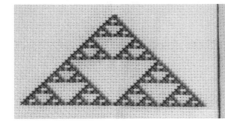

Figure 41. Two realizations of the pattern shown in Figure 40, in the listed colors (right) and in alternate colors (left). The photograph is in focus, despite appearances to the contrary.

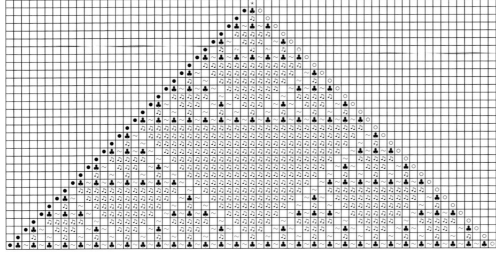

Figure 42. Rule 182's pattern. See Figure 43.

* DMC 561 – Very Dark Jade
● DMC 841 – Light Beige Brown
♣ DMC 838 – Very Dark Beige Brown
○ DMC 524 – Very Light Fern Green
♫ DMC 778 – Very Light Antique Mauve
~ DMC 355 – Dark Terra Cotta

Figure 43. The Rule 182 pattern worked.

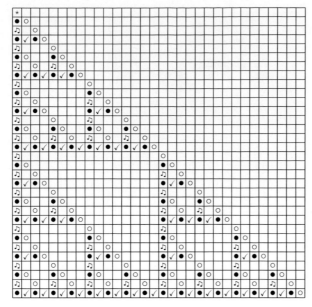

Figure 44. Pattern for Rule 60.

★ DMC 827 – Very Light Blue
○ DMC 813 – Light Blue
● DMC 824 – Very Dark Blue
♫ DMC 825 – Dark Blue
✓ DMC 826 – Medium Blue

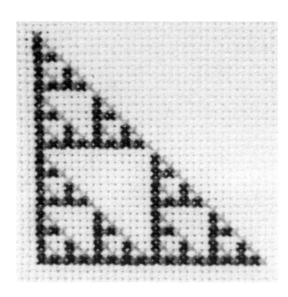

Figure 45. A realization of the pattern shown in Figure 44, stitched in alternate colors.

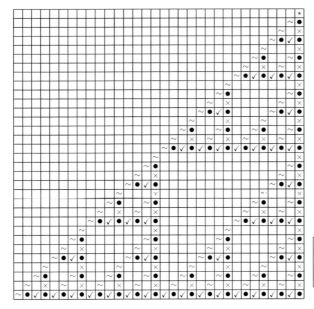

Figure 46. Pattern for Rule 102.

★ DMC 827 – Very Light Blue
~ DMC 813 – Light Blue
● DMC 824 – Very Dark Blue
✕ DMC 825 – Dark Blue
✓ DMC 826 – Medium Blue

Figure 47. A realization of the pattern shown in Figure 46.

Bibliography

[1] Barnsley, Michael Fielding. *Fractals Everywhere*. Academic Press, Cambridge, MA, 1993.

[2] Bowler, Nathan. "Cut-the-Knot Guest book Topic #282." *Interactive Mathematics Miscellany and Puzzles*. Available at http://www.cut-the-knot.org/htdocs/dcforum/DCForumID5/282.shtml, September 2003.

[3] Devaney, Robert L. *A First Course in Chaotic Dynamical Systems: Theory and Experiment*. Addison-Wesley, Reading, MA, 1992.

[4] Edgar, Gerald A. *Measure, Topology, and Fractal Geometry*, Second Edition. Springer-Verlag, New York, 2008.

[5] Falconer, Kenneth. *Fractal Geometry: Mathematical Foundations and Applications*. John Wiley & Sons, New York, 1990.

[6] Hardy, G. H. *A Mathematician's Apology*. The University Press, Cambridge, UK, 1940.

[7] Hensel, Alan. "Conway's Game of Life." *Ibiblio*. Available at http://www.ibiblio.org/lifepatterns/, 2009.

[8] "Conway's Game of Life." *LifeWiki*. Available at http://conwaylife.com/wiki/index.php?title=Conway's_Game_of_Life&oldid=9641, 2009.

[9] McGuire, Michael. *An Eye for Fractals*. Addison-Wesley, Redwood City, CA, 1991.

[10] Rodgers, Tammy M. "Hiding Ends Lily's Way." Available at http://www.frontiernet.net/~Tammy Rodgers/lilys_way.html, 2001.

[11] Schroeder, Manfred. *Fractals, Chaos, Power Laws: Minutes from an Infinite Paradise*, First edition. W. H. Freeman and Company, New York, 1991.

[12] Trevorrow, Andrew and Rokicki, Tomas. "Golly." Available at http://golly.sourceforge.net/, 2009. Open source, cross-platform application for exploring cellular automata.

[13] Voolich, Johanna, and Devaney, Robert L. "The Chaos Game." *The Dynamical Systems and Technology Project at Boston University*. Available at http://math.bu.edu/DYSYS/applets/chaos-game.html, 2010.

[14] Vuilleumier, Bernard. "Chaotic Itinerary but Regular Pattern." *Wolfram Demonstrations Project*. Available at http://demonstrations.wolfram.com/ChaoticItineraryButRegularPattern/, 2010.

[15] Weisstein, Eric W. "Elementary Cellular Automaton." *MathWorld—A Wolfram Web Resource*. Available at http://mathworld.wolfram.com/ElementaryCellularAutomaton.html, 2010.

[16] Wildstrom, D. Jacob. "The Sierpinski Variations: Self-Similar Crochet." In *Making Mathematics with Needlework: Ten Papers and Ten Projects*, edited by sarah-marie belcastro and Carolyn Yackel, pp. 41–53. A K Peters, Wellesley, MA, 2008.

[17] Wolfram, Stephen. *A New Kind of Science*. Wolfram Media, Champaign, IL, 2002.

[18] Wolfram, Stephen. "Pascal's Triangle Mod k." *Wolfram Demonstrations Project*. Available at http://demonstrations.wolfram.com/PascalsTriangleModK/, 2010.

CHAPTER 5

diaper patterns in needlepoint

DIANE HERRMANN

1   Overview

Whenever the subject of diaper patterns comes up, people wonder what this has to do with baby diapers. Historically, diaper patterns were printed on or woven into fabric, and such fabric was used to diaper babies. The use of "diaper" to describe fabric made with such patterns is our baby connection.

The word *diaper* is an ornamentation term that refers to any multicolor pattern of repeating geometric shapes, and is used to describe motifs on pottery, mosaics on walls or floors, or architectural details. Take a look at the curtains, rug, or furniture in the room where you're reading this book. Usually at least one of these will display a diaper pattern. Repeating shapes woven into drapery, square tiles on a floor, and repeating areas on a carpet are diaper patterns. Not surprisingly, patterns on wallpaper can also be diaper patterns.

In the context of needlepoint, diaper patterns have a more restricted meaning, as we will explain shortly. There is a mathematical field of study, crystallographic groups, that includes the study of wallpaper patterns and their symmetries; needlepoint diaper patterns fall in this same mathematical area. As we will see, the two types of study overlap but neither contains the other. In this chapter, we will explore diaper patterns created in needlepoint, and investigate their mathematical properties.

1.1   Needlepoint

Historically, needlepoint was the name given any embroidery done on canvas, using wool to execute a particular stitch (the tent stitch). *Canvas* is material made of stiffened linen, loosely woven in a square grid, identified by count, i.e., how many squares there are per inch. The *tent stitch* is simply one diagonal stitch taken over one intersection of a vertical and a horizontal canvas thread. Often needlepoint patterns appear on dining room chair seat cushions. My own mother worked on one of these tent-stitch cushions for years! (See Figure 1.)

Figure 1. My mother's needlepoint chair cushion.

Contemporary needlepoint (or canvas embroidery), as defined by the American Needlepoint Guild (ANG), is any counted or free stitchery worked by hand with a *threaded* needle on a readily countable ground. Thus, crewel embroidery, cross-stitch, and blackwork are all considered types of needlepoint, while crochet and knitting are not. (Neither crochet hooks nor knitting needles have eyes to thread.) However, needlepoint usually refers more specifically to decorative stitching on canvas, and the inherent square grid of the canvas restricts the mathematical structure of needlepoint.

While counted cross-stitchers are constrained to working with a single type of stitch, needlepoint artists have an amazing variety of stitches to use. Beginners often work needlepoint samplers so that they can experiment with stitches and effects, several of which are shown in Figure 2. To become a Master Craftsman in canvas embroidery in a guild such as the ANG, one requirement is to create a design that uses diaper patterns.

So what is a diaper pattern in needlepoint? We look first at the classic needlepoint reference by Lantz [2], who describes her work as "a celebration of the square

Figure 2. Textured needlepoint stitches.

and its marvels." In her book Lantz defines *Diaper* to be

> the medieval word for a repeating textile pattern, usually diamond- or lozenge-shaped. This scheme of weaving warp and woof, causing reflections of light to dance on its surface, gave it subtle visual mystery. Subsequently, the term was applied to heraldic ornamental sectioning, later still to all small scale geometric repeat patterns that, irrespective of material, related to the diamond and the square. Often interspersed by parallel lines, the fields left by these diagonal, horizontal and vertical divisors may be further decorated by varied motifs. [The word] Diaper is Greek-based, usually thought to mean, "white at intervals" [2, p. 21].

In modern terms, Strite-Kurz explains that "a diaper is a unit of design, which when repeated enough times, forms a visual diagonal in both directions" [5, p. 3]." Mathematicians will recognize that this description specifies that the design have two independent translation axes, which is the underlying requirement for a mathematical wallpaper pattern.

In Section 2, we discuss wallpaper patterns in general and then carefully define needlepoint diaper pat-terns. We also compare and contrast the symmetry groups of these types of patterns. Then, in Section 3, we discuss some ways that one might use needlepoint patterns to identify symmetries. Even young students working on plastic canvas with yarn can make simple di-aper patterns, and more advanced students can take on the design challenge of creating all 12 possible symme-try types with diaper patterns. Section 4 gives detailed instructions for a Symmetry Cube.

2   Mathematics

2.1   Wallpaper Patterns

If a design admits translational symmetry in only one direction (and its opposite direction), we call it one-dimensional, or a frieze pattern. If a design admits trans-lational symmetry in two or more directions, we call it two-dimensional, or a wallpaper pattern. We will start our analysis of wallpaper pattern symmetries with the tools of rigid motions (isometries) of the plane. No matter how complex they seem, all rigid motions of the plane come from just four actions:

* translation,

* reflection over a line in the plane,

* rotation about a point in the plane, and

* glide reflection.

To work with wallpaper patterns, we must be able to recognize each of these motions, and decide if a given pattern allows each of them. This sounds straightforward, but in practice, it can actually be quite tricky. How will we keep track of the different combinations of symmetry motions and identify which is which? And how many of these patterns are really different? To differentiate pattern types, we identify which combinations of the rigid motions of the plane a pattern admits. In fact, different repeating designs may have the same symmetry group; even though they look different, they will have the same pattern type. When we use the word *pattern*, we mean a pattern type classified by the symmetry group, rather than the actual pattern itself.

Theorem 1 [7] *There are exactly 17 possible wallpaper patterns.*

This means there are only 17 different combinations of symmetry motions that result in distinct patterns. Theorem 1 ignores color issues in its classification, but we will address these in Section 2.3.

Mathematicians refer to the 17 wallpaper patterns listed in Table 1 via notation established by the International Union of Crystallography (ICU), which identifies the particular symmetries of a pattern by a sequence of symbols. Understanding this notation will help us explore diaper patterns in a mathematical context.

Following [3], our approach to this notation will be geometric. Because two-dimensional patterns tessellate the plane, each will have an underlying lattice of points that defines the minimum polygonal region repeated by

Type (Full)	Type (Short)	Lattice Type	Rotation Order	Reflection Axes
p1	p1	parallelogram	none	none
p211	p2	parallelogram	2	none
p1m1	pm	rectangle	none	parallel
p1g1	pg	rectangle	none	none
c1m1	cm	rhombus	none	parallel
p2mm	pmm	rectangle	2	90°
p2mg	pmg	rectangle	2	parallel
p2gg	pgg	rectangle	2	none
c2mm	cmm	rhombus	2	90°
p4	p4	square	4	none
p4mm	p4m	square	4	45°
p4gm	p4g	square	4	90°
p3	p3	hexagon	3	none
p31m	p31m	hexagon	3	60°
p3m1	p3m1	hexagon	6	none
p6	p6	hexagon	6	none
p6mm	p6m	hexagon	6	30°

Table 1. The 17 wallpaper groups and their ICU notation.

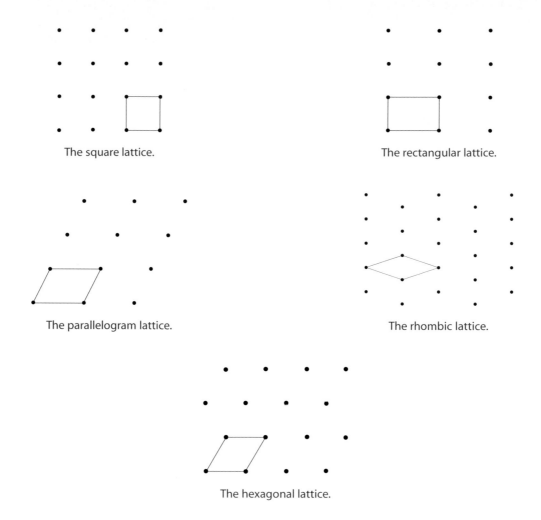

The square lattice.

The rectangular lattice.

The parallelogram lattice.

The rhombic lattice.

The hexagonal lattice.

Figure 3. The five lattices for the 17 wallpaper groups.

translation. There are five such lattices for the 17 wallpaper groups, shown in Figure 3.

In each case, if we form a parallelogram by connecting lattice points and allow no additional lattice points on the edges or in the interior of this parallelogram, we will have identified the part of the pattern which, when translated, will tessellate the plane. Such a parallelogram is called a *primitive cell* for the pattern. These primitive cells are of five types: parallelogram, rectangular, rhombic, square, or hexagonal. The primitive cell in the hexagonal lattice consists of a rhombus made from two adjacent equilateral triangles, and six of these cells surround every lattice point.

In all but one of these five cases, the reflection axes for the pattern are along the edges of these primitive cells. This is not the case for the rhombic lattice. So in this lattice, crystallographers enlarged the rhombic primitive cell to a rectangle twice the size of the primitive cell; with this modification, the reflection axes for this lattice are then along the edges of this rectangle.

This last cell, with the rhombic primitive cell at its center, is called a *centered* cell. See Figure 4.

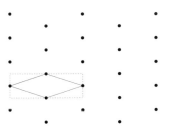

Figure 4. The rectangle for the rhombic centered cell.

At the vertices of each of these primitive or centered cells, we will find the highest order of rotation of the pattern. Since some patterns admit more than one kind of rotational symmetry, we look for the smallest angle through which a rotation preserves the pattern. We then use that angle to identify the pattern type. For example, if a pattern admits a 90° rotational symmetry, it will also admit a rotational symmetry of 180°; we use the fourfold symmetry rather than the twofold symmetry to describe such a pattern.

The full ICU notation consists of a four-symbol sequence we will denote $\alpha\beta\gamma\delta$, and includes information about the cell type, the highest order of rotation, and nontranslation symmetry axes in two directions. For α, we use the letter *p* or *c* to indicate whether the pattern has a primitive or a centered cell. For β, we use an integer to denote the highest order of rotation. Because of the geometry of the cells, the only possibilities for β are 1 (no rotational symmetry), 2, 3, 4, or 6.

The next two symbols characterize the mirrors and glides relative to one translation axis of the pattern. To visualize this, let the left edge of the repeated cell be the *y*-axis. The symbol γ denotes what kind of symmetry axis is perpendicular to this *y*-axis. The letter *m* indicates a mirror axis (or reflection), the letter *g* indicates a glide reflection axis, and the number 1 indicates there is no

perpendicular symmetry axis of either type. The symbol δ denotes an additional symmetry axis. The absence of a symbol in the third or fourth position means the pattern admits neither mirror nor glide reflections.

Crystallographers have shortened some of these designations by deleting symbols when symmetries can be deduced from remaining symbols, as long as there is no confusion with other patterns. For example, *p4mm* can be shortened to *p4m*, since the presence of fourfold rotational symmetry along with mirror symmetry perpendicular to the *y*-axis implies mirror symmetry at a 45° angle to each translation axis.

Faced with a repeating pattern, how do we decide which of the 17 types it is? To see how this all works with a specific example, consider the floral pattern in Figure 5.

Figure 5. A floral wallpaper pattern.

We first identify the lattice that defines the translations of the pattern. The pattern has translation symmetry in two perpendicular diagonal directions (and also horizontal and vertical directions). There are many choices about where to locate the translation axes in this pattern, and Figure 6 shows one such choice. Consider one of the squares formed by the intersecting lattice lines. This is the square cell or *translation unit* for the pattern.

Figure 7 shows this translation unit together with the *fundamental domain*, which is the smallest part of the design that can be rotated and reflected to create the translation unit.

Figure 6. Translation axes determine the lattice underlying the floral wallpaper pattern from Figure 5.

There are different possible choices for selecting a fundamental domain, but lines of reflection must form edges of the fundamental domain.

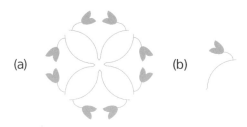

(a)   (b)

Figure 7. The (a) translation unit and (b) fundamental domain for the floral wallpaper pattern from Figure 5.

Once we have identified the translation unit, we check it for centers of rotation and determine the corresponding orders of rotation; the only possibilities are twofold, threefold, fourfold, sixfold, or none. Identifying the order of rotation of the pattern narrows the choices in Table 1. There are two types of centers of rotation in our floral example. There is a fourfold center of rotation in the center of each cell and a twofold center of rotation on the midpoint of each cell edge.

Next, we check to see if there are any reflections or glide reflections in the pattern. If neither is present, the pattern is denoted by only two symbols. After determining which reflections or glide reflections are present, we match our information with the data in Table 1. The floral pattern has *p* for the primitive (square) cell, the number 4 for the 90° rotation, and two *m*'s for the two perpendicular reflection axes. This pattern is thus of type *p4mm*, or *p4m* in the shortened form.

A useful flow chart to help identify the symmetry pattern of any one-color, two-dimensional design appears in [6, p.128]. Mathematically sophisticated readers may want to consult [6] and [3] for a more detailed look at identification.

In [4], Shepherd establishes which of the 17 wallpaper patterns are realizable in cross-stitch.

Theorem 2 [4] *There are exactly 12 cross-stitch realizable wallpaper patterns, listed as the first 12 in Table 1.*

Because there are more needlepoint stitches than the basic cross-stitch available to needlepoint designers, it might seem that we could also realize more than 12 of the 17 patterns in needlepoint. We note that because needlepoint (like cross-stitch) is stitched on fabric with an inherent square grid, symmetry transformations of needlepoint patterns are thus confined to those possible on this square grid. It follows that for needlepoint patterns, we have the following, based on Theorem 2:

Theorem 3 *There are exactly 12 wallpaper patterns that can be realized as needlepoint patterns, listed as the first 12 in Table 1.*

Notice that the rectangular, rhombus, and parallelogram lattices can be superimposed on a finer square grid. While hexagonal (or triangular) repeating patterns may be mimicked in needlepoint, the hexagons cannot be regular. A true hexagonal pattern would require hexagonal ground canvas, and none is currently manufactured that is strong enough to support needlepoint stitches.

2.2 Defining Diaper Patterns

The definition of a diaper pattern should ensure compliance with needlepoint guild standards. These forbid

* *powdering*, or scattering of individual motifs on a background,

* *groundings*, or monochromatic stitches that make only a textural pattern (see Figure 8),

Figure 8. Groundings are not considered diaper patterns.

* *stripes* in any single direction, both actual and those visually suggested (see Figure 9 and the ensuing discussion).

Additionally, a diaper pattern must include visual diagonals as discussed in Section 1.1.

Definition 2 A *needlepoint diaper pattern* satisfies the following conditions.

1. It must extend to a tessellation of the plane with translational symmetry in at least two directions (and thus must also be a wallpaper pattern as defined in Section 2.1). This leads the viewer's eye in two independent directions.

2. The cell must be based on a parallelogram, square, rhombus, or rectangle lattice (and thus must be stitchable as per Theorem 2).

3. The translation unit must include at least two colors. (Navy blue and sky blue would satisfy this condition, even though they are both blues; they have sufficiently different values that the viewer can distinguish them.) This prevents groundings.

4. There can be no continuous monochromatic path from a point on one edge of one translation unit to the corresponding point on the edge of the adjacent unit. A path that is incident to or passes through a vertex where colors meet is not considered monochromatic, as it touches multiple colors there. Alternatively, every continuous path from a boundary point of the translation unit to the corresponding boundary point on the opposite edge of the translation unit must either cross a color boundary line or pass through a vertex where colors meet. See Figures 9 and 10.

Figure 9. Zigzagging chevrons are not diaper patterns because they visually suggest stripes.

The fourth condition in the diaper pattern definition ensures that the pattern will not have stripes. The chevron pattern in Figure 9 has visual stripes.

Here, the translation unit is a two-color parallelogram, as shown in Figure 10; because we can draw a continuous path in the red area, this violates the fourth condition. Note that the path does not cross a color boundary as it traverses the cell. Despite its violation of the diaper pattern definition, this pattern is a legitimate wallpaper pattern.

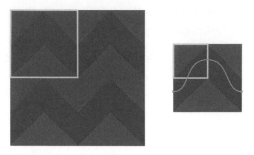

Figure 10. The translation unit and fundamental domain for the chevron pattern.

In contrast, consider the stripeless standard checkerboard. The translation unit is a two-color diamond, as shown in Figure 11.

A continuous path from one cell to an adjacent cell may appear to be monochromatic, but must cross a color vertex on the edge of a cell, rendering it bicolored.

Figure 12. Diaper networks based on tessellating polygons: (a) square, (b) brick, (c) triangle, (d) diamond, and (e) half-drop.

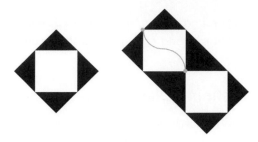

Figure 11. The translation unit and bichromatic path for the standard checkerboard.

To create a guild-approved needlepoint diaper pattern design, the needle artist must work within one of the networks shown in Figures 12 and 13, and must select stitches and colors that will make the visual diagonals apparent.

The reader is invited to verify the ICU designations for the symmetry types of the networks shown in Figure 12: the square and diamond networks are of type *p4m*, the triangle network is of type *pmm*, and the brick and half-drop networks are of type *cmm*.

Although the scale and ogee networks of Figure 13 appear to have curved boundary lines, these networks only approximate the curves with line segments. Notice that the scale network is of type *cm*, whereas the ogee network is of type *cmm*.

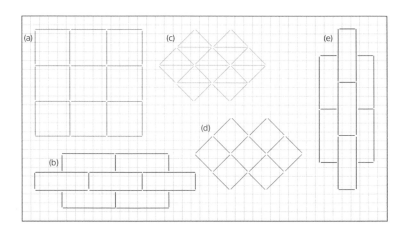

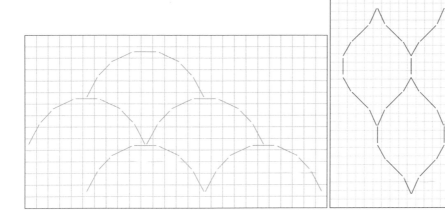

Figure 13. Scale and ogee networks.

Figure 14 shows a schematic for a square-based repeating unit that could be used to make a needlepoint diaper pattern on canvas. Its symmetry type as a line drawing is *p2mm*, because it has twofold centers of rotation and perpendicular lines of reflection. In Figure 15, there are three stitched samples of this design, executed in different colors. Notice how different elements of the pattern appear more or less prominent, depending on the color used to stitch the design. For example, the purple sample has more prominent visual diagonals than the green or pink samples.

Figure 14. Diaper pattern graphed for tent stitch.

Figure 15. Pink, green, and purple versions of the pattern in Figure 14.

2.3 Identifying Symmetries in Diaper Patterns

How do we decide the symmetry type of a diaper pattern? We start with two simplifying assumptions: every translation unit will be completely stitched (no blank canvas) and we will only consider diaper patterns with exactly two colors. Then, we will use three levels of gradation.

1. On the coarsest level, consider only the outline of the stitched design and determine the symmetry type of its underlying network.

2. Next, consider also the color pattern, and determine the effect of any additional restrictions on the symmetry type.

3. At the finest level, consider also the pattern made by stitch orientation and layering, and determine the effect of any additional restrictions on the symmetry type.

In Section 2.2 we determined the symmetry types of the sanctioned diaper networks. We will discuss the inclusion of color here and the effect of texture in Section 2.4.

Considering color in addition to symmetry complicates the classification of symmetry type more than one might think. What exactly is a two-color pattern and how is it different from a one-color pattern? A coloring of the plane that uses exactly two colors will be called a *two-color pattern* if there is some rigid motion of the plane that interchanges the colors. Note that when we consider color, the translation unit is usually different (and larger). For example, a checkerboard has two colors. If we ignore color, the translation unit is a single monochromatic square. However, if we consider color, the translation unit is twice as large because it must encompass the area of two squares of different colors, as pictured in Figure 11.

Two-color symmetry patterns are discussed in detail in Chapter 6, Section 2.6. There the author lists all 46 two-color patterns, specifies the 40 that are available for needlepoint patterns, and also gives extended ICU notation for these patterns. We will use that notation in the subsequent text.

Let's examine one of the simplest wallpaper patterns, $p1$, which allows only translations. Figure 16 shows two versions of a pattern with an asymmetric translation unit, each in two colors, yellow and black.

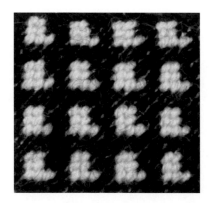 

Figure 16. The asymmetric translation unit as a one-color (left) and as a two-color (right) pattern.

In both examples, the underlying network is square. However, the two patterns have been colored differently. One has the asymmetric motif in yellow on a black background. The other uses alternating black and yellow asymmetric motifs for the repeated design. Both are wallpaper patterns of type $p1$. The first pattern does not allow any symmetric motions that reverse the colors, while the second pattern does. In the second pattern, the asymmetric motifs alternate color in both the horizontal and vertical directions. Notice that the underlying network has larger squares than in the one-color scenario. This design has type $p1/p1$ (see page 129 in Chapter 6). Note that the first pattern could be considered powdering, since it consists of a motif on a background. The all-yellow-motif pattern is therefore neither a true two-color pattern nor a legitimate diaper pattern.

In a different example, consider the chevron stripe in Figure 9. If instead of that arrangement of colors, we colored the units as in Figure 17,

Figure 17. A chevron-like diaper pattern.

we would now have a diaper pattern. This coloring of the wallpaper pattern does not have the visual stripes that the standard chevron pattern does.

2.4 Stitch Texture: Orientation and Layering

By working with the example of the checkerboard, we will see how stitch orientation and layering affect the symmetry type of a pattern. Considering these added design elements raises the possibility that some of the symmetries of the underlying square grid will no longer preserve the pattern.

There are several standard ways to fill a square region using needlepoint stitches; stitching directions for some of these are given in Section 4. We first consider eyelet stitch, which consists of straight stitches arranged so that they meet in a center hole in the canvas. A two-color checkerboard filled with eyelet stitch is shown in Figure 18.

Figure 18. A small eyelet checkerboard.

Whether we consider an eyelet checkerboard as a one-color pattern or as a two-color pattern, it retains all of the symmetries of the underlying checkerboard and so has symmetry type $p4m$.

A second standard stitch is the Gobelin stitch, where straight stitches of equal length are worked vertically, horizontally, or diagonally. A two-color checkerboard filled with Gobelin stitch is shown in Figure 19. Notice that the centers of rotation are no longer fourfold, but instead twofold, and that there are vertical and horizontal mirrors but not diagonal mirrors. Hence, the symmetry type of the Gobelin stitched checkerboard is pmm. By adding stitch orientation to symmetry considerations, we see that the symmetry type changed from that of the underlying network.

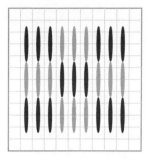

Figure 19. Checkerboard in two colors, Gobelin stitch version.

A third standard stitch is the Scotch stitch, which fills a square region with five diagonal stitches. A two-color checkerboard filled with Scotch stitch is shown in Figure 20.

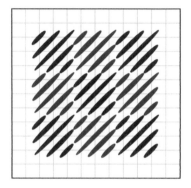

Figure 20. Checkerboard in two colors, Scotch stitch version.

As with the Gobelin stitch, the Scotch stitch has twofold rotational symmetry. Note that the two reflection axes are on the diagonals through the centers of the squares, in perpendicular directions. Because of the stitch orientation, there are neither horizontal nor vertical axes of reflection. With stitch orientation taken into account, the pattern has symmetry type *pmm*, not the same type (*p4m*) as the underlying square grid.

You might be wondering why these two stitched checkerboards, one done in Gobelin stitch and one done in Scotch stitch, have the same symmetry type. Both versions are executed using straight stitches, and in each version, adjacent squares include only parallel stitches. These common characteristics lead to the same symmetry patterns, even though different stitch lengths are used to fill the squares.

Can we fill a checkerboard with needlepoint stitches and display a pattern that has a symmetry type other than *p4m* or *pmm*? A stitch called the woven cross stitch will do the trick. This stitch, shown in Figure 21, consists of four interlaced diagonal stitches.

Figure 21. Checkerboard in two colors, woven cross stitch version.

(For a closeup of this stitch, refer to Figure 27 in Section 4.) The center of the translation unit is a fourfold center of rotation, but the woven cross stitch has no reflection symmetry because of its interlacing. Thus, a checkerboard stitched with woven cross stitch has symmetry type *p4*.

3 Teaching Ideas

The study of geometry is intertwined with the study of symmetry from the elementary grades through college courses. Likewise, the study of symmetry can be elementary or advanced, depending on the student. Elementary school students can point out and locate repeating patterns in their classrooms by looking at the floor tiles or classroom posters. Beginning college-level students can begin to classify patterns by the symmetry motions that they allow, and more advanced students can continue with the study of more abstract symmetry groups of patterns.

In almost every classroom, someone will be wearing an article of clothing with a woven or printed repeating pattern. Classic men's ties are almost always examples of the diaper patterns we have described in this chapter. In an elementary school classroom, students could identify which shapes make up the repeating pattern. Middle-school geometry students who have an idea about the four kinds of symmetry motions could discuss whether the pattern on the clothing allows translation, reflection, or rotation of any kind. College students who have been introduced to wallpaper patterns in a liberal arts mathematics class should be able to identify whether the print on the clothing is of type *pmm* or *p4m*, for example.

Creating repeating patterns and using stitch orientation or colors to alter their symmetry types can be explored at a number of levels. Using graph paper and colored pencils to modify simple checkerboard patterns is a good way to start, and increasing the level of sophistication in design will also challenge more advanced students.

Have students mark a square area on graph paper and use a pencil to make a design. Begin by limiting students to diagonal segments over single intersections; an example of such a design is shown in Figure 22.

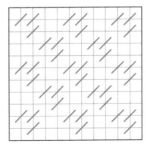

Figure 22. A simple repeating design on graph paper.

You might ask students to identify the symmetries of this design, and then their patterns. Where are the translation axes? What is the translation unit? Where are centers of rotation, and through what angle is the rotation that preserves the pattern? Are there any axes of reflection? Do these axes pass through any centers of rotation? Answering these questions will lead students to identify the symmetry types of their patterns. Comparing the symmetry types that arise across the class can lead to a discussion of what symmetries are preferred by class members.

As a follow up, use motifs in alternating colors to make similar diagrams, and discuss whether the result, if stitched, would be a needlepoint diaper pattern. Or, ask students to create patterns that match a particular symmetry type. For example, asking for type *p2* requires that students make a pattern with a 180° rotation but no axes of reflection. The discussion that ensues when students work in pairs to create and analyze patterns is especially fruitful.

Young students might benefit from counting along the canvas grid as required to make stitches of the right length, such as in Gobelin stitch. Elementary school art classes often include projects that use a threaded needle, and plastic canvas with its relatively large grid makes the stitching easy. Craft stores also sell plastic needles with large eyes that young children will be able to thread easily. Working on a two-dimensional grid to make sloped line segments as in Scotch stitch or eyelet stitch is also beneficial. For students learning to plot points on a graph, these kinds of designs can provide a creative outlet for an otherwise routine exercise. Stitching a model of a Scotch stitch will allow students to manipulate their samples to see how the rotation of 180° preserves the stitch orientation. Stitches of different slopes can be used to investigate which stitch patterns preserve the symmetries of a grid and which break those symmetries.

There are many resources for stitch diagrams; [1] and [2] stand out. Consulting these to make designs on plastic canvas from the diagrams in the books can enhance graph-reading skills. Students already familiar with graphs and graph paper will be able to examine stitch diagrams and follow patterns to create complicated stitches. Teachers can look for stitch diagrams that illustrate the kinds of symmetry they want to present in a classroom setting. Interpreting a stitch diagram and using layering to execute complicated stitches also improve a student's understanding of how to construct a three-dimensional design from two-dimensional instructions. Figuring out the steps needed to stitch the pattern in Figure 14 is a good way to understand sequencing of stitches. Such skills will be useful for students who are learning to use the tools of straightedge and compass to construct geometric objects and provide step-by-step proofs of geometric theorems. Once students have progressed to reading these kinds of diagrams, they could create tent-stitched versions of the diagrams they made with colored pencils on graph paper. (Instructions for tent stitch are included in Section 4.)

With a little yarn and some needlepoint or plastic canvas, students can create a Symmetry Cube for use in class. Each face of the cube is stitched in just two colors, and only simple straight and diagonal stitches are needed. Two explorations for students follow.

⋆ Identify the symmetry type on each face of the cube by examining the symmetries of the underlying network, the symmetries of the pattern including stitch orientation, and the symmetries of the pattern including both stitch orientation and color.

⋆ Using graph paper, students can create needlepoint diaper patterns (remember that cross-stitch and blackwork count as needlepoint!). Different parts of a pattern can be worked in different colors to make it easier (or more challenging) to discover its symmetry type. Notice that instead of beginning with a symmetry type and designing a diaper pattern to match, this

activity begins with the design of a diaper pattern and then asks students to identify its symmetry type.

3.1   Project Ideas

The Symmetry Cube has six faces, and when color and stitch orientation are considered, it illustrates six of the 12 allowable symmetry patterns. Create a different symmetry cube with faces that exhibit the other six symmetry types available in needlepoint designs.

Create a symmetry dodecahedron, with all 12 wallpaper patterns included. This project is more complicated than the previous two, but would be a good stretch for students who have enjoyed the earlier projects. The dodecahedron offers a special twist because its faces are regular pentagons. Therefore, design units will not tile the faces without crossing the pentagon edges and stitchers must compensate for that fact by stitching as much of the design as possible in the pentagonal area available. Selecting simple designs and basic stitches will help reduce the need for compensation.

4   Crafting the Symmetry Cube

The cube shown in Figure 23 can be stitched on needlepoint or plastic canvas. The adventurous and advanced stitcher may use cross-stitch fabric, but will need interfacing to form the result into a sufficiently rigid cube. A hint of color theory is included in the arrangement of the primary and secondary colors on the cube; each vertex represents a relationship on the color wheel. The cube has two special vertices—one where the three primary colors meet and one where the three secondary colors meet. The faces have been arranged so that these are opposite. The reader is invited to discover the interesting features of the remaining vertices.

Figure 23. A Symmetry Cube, complete with snake.

Materials

* 10″ by 12″ piece of 13- or 14-count white mono canvas or 7- or 10-count plastic canvas. To prevent mono canvas from fraying or catching on your thread as you are stitching, wrap the edges with masking tape, or sew bias tape around the edges.

* 10″ by 12″ stretcher bars and tacks for mono canvas. Because canvas is made of stiffened threads, it is easily distorted. (It is also much too rigid a material to fit into a hoop.)

* #20 tapestry needle (younger stitchers may do better using a plastic needle with a large eye for easy threading).

* One skein each of DMC pearl cotton #3 in colors of red (#666), orange (#970), yellow (#727), green (#699), blue (#995), and purple (#552). Two skeins of black (#310).

or

Use one or two strands of Paternayan® yarn (100% wool) on mono canvas, or one strand of light-weight knitting yarn with plastic canvas. (Using worsted-weight knitting yarn with plastic canvas creates difficulties when multiple strands must pass through a single hole.)

* Pen or sharpie for marking the cube.

* Polyester fiberfill for stuffing.

Basic Needlepoint Instructions and Tips

Follow each stitch diagram in numbered sequence, coming up at odd numbered places on the diagrams, and bringing the thread to the underside of the canvas at the even numbers. Work with thread that is about 18″–24″ long.

To begin the first thread in the design, use a waste knot (see Figure 24).

Figure 24. In-line waste knot.

About 4″ from the start of your first stitch, insert the needle on the front of the canvas so that the knot is on the front of the canvas and the thread runs behind. Bring the tip of the needle to the front of the fabric to start the first stitch, and stitch the area. Your stitches will be covering the thread that is on the back of the canvas. When you reach the knot on the top of the canvas, or when you finish that row, carefully snip the waste knot from the front of the canvas.

Alternately, you can secure a beginning or ending thread by weaving it through nearby threads on the underside of the stitching. To secure the tail, run your needle under a few threads in one direction, take one or two backstitches over a thread, and then run your needle under a few threads in the opposite direction.

To reduce bulk, try not to begin and end threads under the same stitches. Also, do not weave dark thread under light thread, as this may discolor the front of your piece. Remember that the back of the Symmetry Cube doesn't have to look pretty—only you will see it.

Instructions for Marking and Stitching the Faces

Use a sharpie or a pen to mark six squares on the canvas in a cross-shaped net (see Figure 25) that will form a cube. You will need to count carefully so that all your squares are the same size, and so that you center the net on the canvas. In the stitched model, the squares' edges are 24 canvas threads; be sure to count 24 threads, and not 24 canvas holes, when you mark the canvas. Begin the top left corner of the first square 2.5″ from the top edge of the canvas, and 4.5″ from the left edge of the canvas.

You can alter the size of your cube by making the squares a uniform, but different, size than the model. If you enlarge the sizes of the squares, you will need to work on a larger piece of mono canvas. If you are using plastic canvas, you need not use the net but instead could make six individual squares, and then cut them out and stitch them together.

The stitch patterns for the faces are given in Figures 26–35. One orientation of each face is shown in Figure 25. However, which side of any face is "up" does not matter, so feel free to rotate the canvas as you see fit. As you stitch near the edge of each face, you may not be able to stitch a complete unit of the design and will need to compensate for that fact by stitching as much of the design as possible in the square area available. You may need to use some partial stitches, but the design idea will still be complete. This process is called *compensation*. In most cases, you will want to stitch all the units of the same color on each face first, then fill in the remaining area with the black yarn.

Red and black face: Scotch stitch. To make a Scotch stitch, begin with a short stitch that crosses just one canvas intersection, called a 1 × 1 stitch. Follow this with a 2 × 2 stitch, a 3 × 3 stitch, another 2 × 2 stitch, and finally another 1 × 1 stitch, as shown in Figure 26.

Figure 25. Net for unassembled Symmetry Cube.

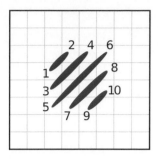

Figure 26. Scotch stitch diagram.

To complete the face, work one diagonal row of Scotch stitches in a single color, starting at the top left corner of the square face and stitching to the bottom right corner of that square face. When you have completed a diagonal row in one color, thread your needle with the other color and work another diagonal row above or below the first. Continue until the square is filled. The pattern will resemble that of Figure 20 in Section 2.4.

Purple and black face: woven cross stitch. Stitching instructions for a single woven cross stitch are given in Figure 27.

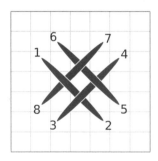

Figure 27. Woven cross stitch diagram.

Be sure to slip the last stitch of each woven cross under the first stitch of that woven cross. To complete the face, work one diagonal row of woven cross stitches in a single color, starting at the top left corner of the square face and stitching to the bottom right corner of that square

face. When you have completed a diagonal row in one color, thread your needle with the other color and work another diagonal row above or below the first. Continue until the square is filled. The pattern will resemble that of Figure 21.

Orange and black face: criss cross Hungarian diagonal stitch. Work stitches from left to right across the canvas in the pattern indicated in Figure 28.

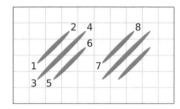

Figure 28. Criss cross Hungarian stitch diagram.

Notice that stitch 7 comes up in the same row as, and four canvas threads to the right of, stitch 1. When you end the first row, change to black yarn and work from right to left to fill in the second row in the same way you worked the orange. Look at Figure 29 to see where to begin the second row of stitches. Continue these rows until you reach the bottom of the square. At the edges of the square, you will need to compensate as shown in Figure 29 so that the design ends appropriately in each row.

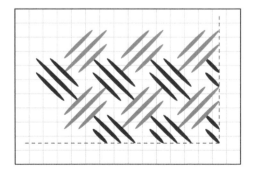

Figure 29. Partial pattern for orange and black face.

Yellow and black face: nonsymmetric unit repeat. This motif has no internal symmetry, so copies of it will be stitched in alternating colors of yellow and black, with black and yellow backgrounds (respectively) as shown in Figure 30 in order to make this a two-color pattern.

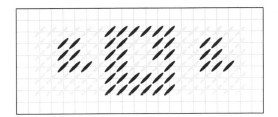

Figure 30. Two-color asymmetric motif stitch diagram.

Tent stitch will be used for both the motifs and the backgrounds. The tent stitch is just a diagonal stitch over a single canvas intersection. The direction of stitching determines in what order you place your stitches. If you are stitching from right to left, use Figure 31; if you are stitching from left to right, use Figure 32.

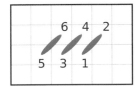

Figure 31. Tent stitch diagram, right to left.

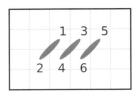

Figure 32. Tent stitch diagram, left to right.

While the stitch order shown may not seem logical, it will insure good coverage of the canvas and will also keep it from distorting.

Green and black face: leaf stitch. Figure 33 indicates (by odd numbers) the places to bring the needle up in stitching a leaf. The first stitch is a vertical stitch over three canvas threads; notice that the second and fifth stitches go down in the same hole where you ended the first stitch. The third and fourth stitches both end one canvas thread below where the first two ended.

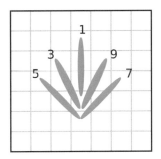

Figure 33. Leaf stitch diagram.

Start by stitching a row of green leaf stitches across the square left to right. This stitch works best if you start at the bottom of the square and add rows on top of those just stitched. Begin a row of black leaf stitches on the left above the green leaf stitches, as shown in Figure 34.

Figure 34. Partial pattern for green and black face.

You have two choices for working the partial leaf stitches on the edges. One way is to work all the complete leaf stitches in the row, and then come back to put in the partial stitches. Alternatively, you could start on the edge with a partial leaf stitch to compensate at the edge of the square. With either method, continue to stitch alternate rows of green and black leaf stitches until you fill the square.

Blue and black face: diagonal triangles. These two-color triangles are made with diagonal stitches. Beginning at a corner of the canvas face, follow the numbers given in Figure 35 to complete the diagonal black stitches. Then, change to blue yarn and begin at A; complete the stitch at B and thus cover the ends of the black stitches.

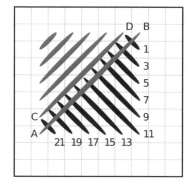

Figure 35. Diagonal triangle stitch diagram.

Continue to C, and do all six of the blue stitches. Continue with the next set of black stitches, and move across the rows to fill in the square. It will help in counting to see that the first black stitch of each unit begins six threads to the right and one thread below the left corner of the unit.

Finishing

Canvas. When you have finished stitching all six squares, machine stitch around the edge of the cross-shaped net so that the edges are secure. If you don't have a machine, you will need to be careful as you assemble the cube so that the cut canvas threads don't unravel. Next, backstitch around the cube's edges in black yarn. You don't want to leave an uncovered canvas thread between the backstitches and the faces. So, execute these backstitches in the same holes as the already stitched edges of the faces. Cut out the entire cross-shaped design, leaving at least five unstitched canvas threads on each edge to make a flap, as shown in Figure 36. Clip the corners diagonally, all the way to the machine stitching. Fold in the excess canvas; you might want to tack the folded canvas to the front with straight pins or basting thread to make it easier to work with. You will sew two edges together by lacing yarn through the backstitches on the edge of the cube. If you start at a corner, these backstitches should match up. Before you close the last seam, stuff the cube with the polyester stuffing.

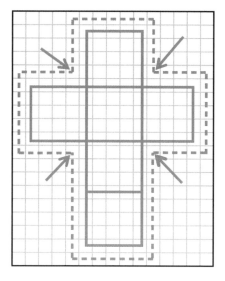

Figure 36. Cutting out the cube.

Plastic. Cut out each of your squares carefully in the row just outside your stitching, making sure not to cut through any of the stitching. Whip stitch the squares together with black sewing thread and a smaller needle. If you want to preserve the color relationships of the vertices, assemble the cube with faces in relationship as they are in the net (Figure 25). Before you close the last seam, stuff the cube with the polyester stuffing.

Once the cube is assembled, use one strand of black yarn to overcast the edges.

Bibliography

[1] Christensen, Jo Ippolito. *The Needlepoint Book.* Simon & Schuster, New York, 1999.

[2] Lantz, Sherlee. *A Pageant of Pattern for Needlepoint Canvas.* Grosset & Dunlap, New York, 1973.

[3] Schattschneider, Doris. "The Plane Symmetry Groups: Their Recognition and Notation." *American Mathematical Monthly* 85 (1978), 439–450.

[4] Shepherd, Mary. "Symmetry Patterns in Cross-Stitch." In *Making Mathematics with Needlework: Ten Papers and Ten Projects*, edited by sarah-marie belcastro and Carolyn Yackel, pp. 71–89. A K Peters, Wellesley, MA, 2008.

[5] Strite-Kurz, Ann. *Diaper Patterns.* Strite-Kurz, Self-published, 2007.

[6] Washburn, Dorothy K. and Crowe, Donald W. *Symmetries of Culture: Theory and Practice of Plane Pattern Analysis.* University of Washington Press, Seattle, 1991.

[7] Weyl, Hermann. *Symmetry.* Princeton University Press, Princeton, NJ, 1983.

CHAPTER 6

group actions in
cross-stitch

MARY D. SHEPHERD

WITH SARAH-MARIE BELCASTRO AND CAROLYN YACKEL

1 Overview

The impetus for this chapter was the mathematics of cross-stitched rosettes. Shepherd investigated two aspects of these motifs, and in that process, she designed and cross-stitched all pieces shown in this chapter. One aspect led to the use of cross-stitch with abstract algebra students. The other generalized the results in [7] to a larger class of patterns. The theme of group actions unifies these two projects.

Group actions are all around us. If you have ever shuffled cards, spun an open umbrella, or rolled dice, you have experienced a group action. A group is a mathematical object (that we shall define shortly), and its elements can act on some underlying set by moving them around. All mathematicians, including undergraduates, study group theory as part of a more general study of algebra.

Here is how group actions work in the examples just listed. A new deck of cards is ordered by suit and by card number within suit. Shuffling them rearranges the cards into a new order. If a person is particularly good at card shuffling (for example, magicians practice this skill), then she can execute a *perfect shuffle* that exactly interleaves two halves of a deck. This is a group action: it has a predictable outcome of reordering the cards and can be done over and over. Consider now an open umbrella. It has some number of spokes, k, emanating from the center to the outer edge. Twirling the umbrella so that each spoke ends up where some other spoke was (that is, rotating the umbrella by $\frac{2\pi n}{k}$) is a group action. The set here is the different rotational positions of the umbrella, and the action is rotation. Now imagine dice being thrown. Each die has six sides, and a position of the die is determined by which side is up and which side is to the front. The collection of such positions is an underlying set, and picking up a die and rolling it changes the position of the die to another position in the set. Admittedly, rolling is not predictable, so a die

roller cannot control which group element is acting on the set.

To see if you understand the idea, here are some exercises. For each of these examples, can you figure out what the set is and what the group action is?

* Remove and replace the top of a square bin.

* Reposition dancers within formation on a stage.

* Twist one face of a Rubik's^{TM} cube.

* Walk in someone's footprints on the beach.

* Roll a No. 2 wooden pencil through your fingers as you write.

This chapter is about the many ways that group actions are related to cross-stitch patterns. The repetitions and symmetries of the motifs give us a way to visualize algebraic structures. By using separate colors on different motifs, we can see a richer structure.

The Mathematics section is devoted to elaborating on the ideas presented so far. We begin by defining groups in Section 2.1 and group actions in Section 2.2, and show how to visualize group elements in cross-stitch in Section 2.3. We then proceed to describe group actions on cross-stitch patterns in Section 2.3, with dihedral groups acting on rosette patterns as the primary example. This is followed by instantiating the group operation in cross-stitch patterns in Section 6 and using this operation to investigate subgroups and cosets in cross-stitch (Sections 2.4 and 2.5, respectively). The action of a group on certain cosets underlies the creation of two-color patterns, explicated in Section 2.6. Finally, we classify two-color rosette patterns (Section 2.7), frieze patterns (Section 2.8), and wallpaper patterns (Section 2.9). Section 3 contains activities for students of all levels, with some specific to abstract algebra classes. A number of explorations are analytical and others result in physical creations. Finally, in Section 4 we give directions for crafting a counted cross-stitch pillow that includes four two-color wallpaper patterns and one four-color wallpaper pattern.

2 Mathematics

Seasoned mathematicians may find Sections 2.1 and 2.2 elementary, so may wish to begin reading with Section 2.3. Others may be comforted by the review of the definitions of group and group actions.

2.1 What Is a Group?

A *group* is an algebraic structure on a set of objects with an operation that takes two objects and returns a third. (This is called a *binary operation*.) In other words, if a and b are in the set, and $\star$ denotes the operation, then $a \star b = c$ is in the set. The operation $\star$ can be regular multiplication, or regular addition, or something else entirely. Below, we will refer to $a \star b$ as *starring* a and b. (Traditionally it is referred to as multiplication, whether or not the operation corresponds to the usual multiplication.) There are three other basic conditions that must hold for a structure to be considered a group.

* The associative condition allows us to move parentheses around in algebraic expressions. That is, $(a \star b) \star c = a \star (b \star c)$, so that which starring is done first does not matter; what matters is the order in which the elements are written. (Do not confuse this condition with commutativity; $a \star b = b \star a$ is usually not true, because groups are often not commutative. We will return to this notion later.)

* In every group, there is some element that behaves just as 0 does in addition of numbers. If we call the identity e, then $e \star a = a \star e = a$ for any element a of the group. That means that when e is starred with another element, nothing happens—the result is the other element.

* Every element a in a group has another element a^{-1} called its inverse. An inverse element undoes whatever the original element does. In algebraic notation, $a \star a^{-1} = a^{-1} \star a = e$. This undoing idea can be seen in the group of numbers under the operation of addition. Each number x has an inverse $-x$ (which is usually called the additive inverse) such that $x + (-x) = 0$.

It might appear that a group must have at least three elements. In fact, it is only required to have one element (the identity element). Common examples of groups include $\{0\}$, the integers, and the real numbers, each using addition as the operation.

A *subgroup* is a subset of a group that is also a group on its own, using the same operation. Note that it must contain the identity element of the parent group. For example, the even integers form a subgroup of the integers under addition (and notice that 0 is an even number).

2.2 Defining Group Actions

Informally, a group can act on a set. Sometimes a group acts by rearranging the elements of a set, as when one shuffles a deck of cards. Sometimes a group acts on a set of group elements by replacing one group element with another, as when one rotates the spokes of an umbrella. (Here, the set of group elements is represented by the set of rotational positions.) In these cases, we have group actions. Now we shall be more precise and give the formal definition.

The *permutation group* on a set is the collection of all reorderings of the set. (Notice that the group elements are reorderings, and thus movements, which are active, and not orderings, which are static.) For a set S, we denote the permutation group on S by $\pi(S)$. When a group G acts on a set, it does so as a subgroup of $\pi(S)$.

Formally, a *group action* is a map $G \times S \rightarrow S$ defined by $(g, s) \mapsto g(s)$, where $G \subset \pi(S)$. Notice the special case where S is the set underlying G.

In the specific cases of frieze and wallpaper patterns as considered below, the symmetry groups are infinite. Hence, to reduce abstraction it is easiest to consider group actions from the standpoint of the transformation group G acting on the relevant pattern. Indeed, we will typically abandon our formal definition for this convention.

2.3 Groups Acting on Cross-Stitch Patterns

Symmetries of a pattern are transformations that take the pattern directly back onto itself. These are discussed in some detail in [7]. As a brief review, there are four types of symmetry transformations, namely reflection, rotation, translation, and glide reflection. These are easy to demonstrate in counted cross-stitch because cross-stitch is done on a fabric that has a square grid. The basic stitch is × shaped and covers one square. Partial stitches can cover a half of a square when cut along a diagonal.

Figure 1. The group of symmetries of this pattern is *only* the identity element—the pattern has no symmetries.

The symmetries of a cross-stitch pattern form a group. The elements of this group are transformations of the cross-stitch pattern and the operation ⋆ is composition—first performing one transformation, and then performing a second one. (Note for the advanced mathematician: while a group always acts on the set underlying the group itself, this is usually invisible and so

we will not look at the symmetry group in this manner in the coming subsections.)

A pattern with no symmetry would correspond to an example of a group with only the identity element, as shown in Figure 1.

Representations of group elements in cross-stitch. In this section, our goal is to visualize group elements in cross-stitch. Consider the pattern in Figure 2, which has a reflection symmetry.

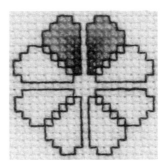

Figure 2. The group corresponding to this pattern has the identity and one reflection symmetry.

If one reflects across the vertical line between the two halves of the figure, the pattern lands directly on top of itself, as is necessary for any symmetry transformation. The group represented by this symmetry pattern has two elements (the identity and one reflection) and is called $\mathbb{Z}_2$. A visual way to represent the elements of the group is to choose a small part of the pattern, such as a single petal in the upper left-hand part of the rosette, and watch what happens to that petal when the different group elements (symmetry transformations) are applied to it. This is shown in Figure 3.

We arbitrarily choose the left-hand rosette to represent the identity element, where no transformation has occurred. The right-hand rosette shows what happens to our chosen petal when the vertical-line reflection is applied to it. It represents the reflection element. Notice that if one does this reflection twice, the result

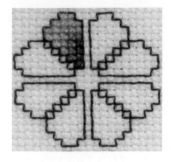 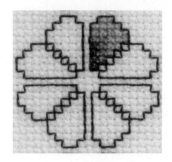

Figure 3. A visualization of the two elements of the group in Figure 2.

is the same as applying the identity transformation (that is, the petal remains unchanged). Thus, repeating the reflection undoes the initial reflection. This means the reflection symmetry has an inverse, namely, itself.

There is mathematical subtlety here: The cross-stitched rosettes shown in Figure 3 are static. Yet, they represent actions, and are in fact the images of those actions on the chosen representation of the identity element. Further, as static objects, each cross-stitched rosette has two contextually different meanings. In this context each rosette represents a group element as just described; outside this context, each rosette has a symmetry group of its own (in this case, simply the identity). That is, the cross-stitched rosettes of Figure 3 are acted on by the symmetry group of the cross-stitched rosette of Figure 2.

We can repeat this analysis for any rosette pattern. Figure 4 shows the rosette pattern with the largest number of symmetries that can be shown in counted cross-stitch. There are four lines of reflection (vertical, horizontal, and two on the diagonals), three rotational symmetries (90°, 180°, and 270°), and the identity. Figure 5 shows the the eight elements of this group, called D_4. Each pictured rosette represents the result of a symmetry transformation applied to the chosen identity element.

Figure 4. The group corresponding to this pattern has four reflection lines, three rotational symmetries, and the identity.

On the other hand, each of the rosettes in Figure 5 could be conceptualized as an individual petal of the rosette in Figure 4.

We can continue this idea with the two transformations that do not appear in the finite (rosette) patterns, namely translations and glide reflections. Figure 6 shows a portion of an infinite wallpaper pattern with type *p4m* (see Section 2.1 of Chapter 5 for details), with the identity element chosen to be the motif outlined in blue. The actions corresponding to the numbered motifs, described in Table 1, represent different symmetry group elements. Each motif in the wallpaper pattern can be reached by a single transformation. Some of these transformations are shown in Table 1.

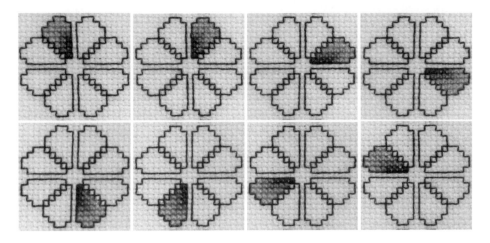

Figure 5. A visualization of the eight elements of D_4.

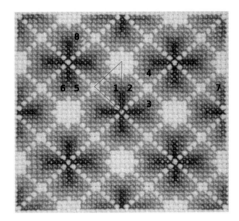

Figure 6. Viewing the elements of the group *p4m* as transformations of some identity motif.

1	Identity
2	Reflect over the vertical line to the right of the motif
3	Rotate 90° clockwise around the lowest point of the motif
4	Rotate 270° clockwise around the highest point of the motif
5	Rotate 180° around the leftmost point of the motif
6	Glide reflect horizontally over the middle of the motif
7	Translate horizontally right
8	Glide reflect over vertical line at the left vertex of the motif

Table 1. Description of the transformations in Figure 6.

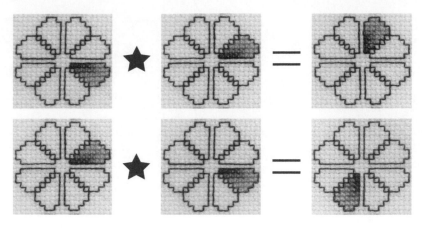

Figure 7. A demonstration that D_4 is not commutative.

The group operation instantiated in cross-stitch. Now that we have depictions of the group elements, we must determine the meaning of the group operation applied to two cross-stitched rosettes. Consider the third column of Figure 5, containing the rosettes that correspond to clockwise rotation by 90° and by 270°. Doing one rotation and then the other—that is, composing these two elements as actions—we rotate by 360° and obtain the identity element. That is, $g_{270°} \star g_{90°} = e$. Notice that we began with static rosettes, moved to their underlying actions, starred those actions, and translated the result to a new static rosette.

In the next example, we see that D_4 is noncommutative. Figure 7 shows that rotating clockwise 90° and then reflecting over the upper-right/lower-left diagonal is not the same as first reflecting over that diagonal and then rotating 90° degrees clockwise.

Note that the transformations shown in Figure 7 are given in the same order as usual functional composition, and therefore the transformations are read (and performed) right-to-left.

Composition of depictions of group elements works similarly in Figure 6. This case is different only in that the wallpaper pattern has infinitely many elements. Any petal motif represents a single group element. That is, any petal motif can be reached by a single group element (symmetry transformation), but there may also be compositions of group elements that result in that motif.

2.4   Subgroups in Cross-Stitch

Recall that a *subgroup* of a group is a subset that is itself a group. Here we will examine subgroups of D_4, although the ideas apply to any group. (Visualizations of general subgroups are explored in [2].) Restricting to an appropriately chosen smaller set of transformations produces a subgroup of the group. We can see this by looking at collections of elements, as with the subgroup shown in Figure 3 and the group shown in Figure 5; or, we can depict a subgroup with a single picture as in Figure 2 with parent group depicted in Figure 4.

All subgroups of D_4 (except for the trivial subgroups $\{e\}$ and D_4) are shown in Figure 8. The top row of Figure 8 shows four two-element subgroups, each formed using the identity element and one of the four reflections. A fifth two-element group is formed using the identity and rotation by 180°. Each of these five subgroups is isomorphic to $\mathbb{Z}_2$. There are two four-element

Figure 8. All the nontrivial subgroups of D_4, where each petal represents a group element.

subgroups; each is comprised of the identity element, two perpendicular reflections, and rotation by 180°, and both are isomorphic to $D_2 \cong \mathbb{Z}_2 \oplus \mathbb{Z}_2$. Finally, there is a four-element subgroup comprised of the identity element and the three rotations. It is cyclic and thus isomorphic to $\mathbb{Z}_4$.

One might notice that some subgroups are also subgroups of other subgroups. For example, the second column of Figure 8 shows a $\mathbb{Z}_2$ subgroup that is also a subgroup of the D_2 subgroup pictured below it. There are six other subgroup-of-subgroup pairs in Figure 8. Can you find them all?

An interesting example of a subgroup is the *center* of a group, which is the collection of all elements that commute with the entire group. The center of D_4 is shown as the first column of Figure 5.

It turns out that a subgroup that is half the size of the original group (called an *index-2 subgroup*) defines a two-color pattern. We discuss this in detail in Section 2.6. There is also an intimate relationship between subgroups and the notion of cosets, which we illuminate shortly.

2.5 Representations of Cosets in Cross-Stitch

For a group G with subgroup H, a *coset* of H in G is a set of the form $a \star H = \{a \star h \mid \text{fixed } a \in G, \text{all } h \in H\}$ (left coset) or $H \star a = \{h \star a \mid \text{fixed } a \in G, \text{all } h \in H\}$ (right coset). Notice that usually a coset will not contain the identity element of the group and so a coset is generally not a subgroup itself. Conceptually, a coset shifts a subgroup by a group element a. It follows (with some work) that cosets partition a group. That is, any two cosets $a \star H$ and $b \star H$ (or $H \star a$ and $H \star b$) of a subgroup H have the same size and are either identical or have no common elements. Further, every element of G is contained in some coset; $a \star H$ contains the element $a \star e = a$.

In order to visualize cosets using cross-stitch, we return to the notion of group actions on elements. We will begin with a depiction of a subgroup H, and act on that subgroup with some group element a, to produce a depiction of the coset $a \star H$ (or of $H \star a$). Because distinct cosets are disjoint, we can use colors to represent the cosets, without risk of confusion from overlapping colors. Additionally, depictions of cosets will cover the entirety of the original pictured group.

We now examine D_4, though this analysis will hold for any cross-stitched symmetry group (and in particular we will examine wallpaper groups in Section 2.6). Consider the two element subgroup H_v determined by vertical reflection. Figure 9 depicts all left and right cosets of the pink cross-stitched subgroup H_v. Consider $g_{90°}$. The coset $g_{90°} \star H_v$ rotates the subgroup H_v by 90°, producing the blue petals in the top rosette of Figure 9; the coset $H_v \star g_{90°}$ applies the identity and vertical reflection actions to the petal representing $g_{90°}$, yielding the two blue petals in the bottom rosette of Figure 9. The other cosets are produced in a similar manner.

Figure 9. Left (above) and right (below) cosets—each coset is a different color.

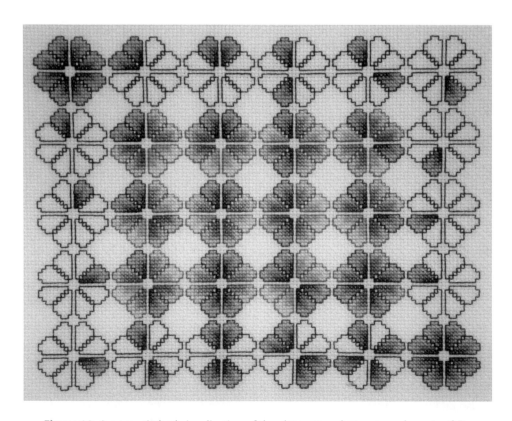

Figure 10. A cross-stitched visualization of the elements, subgroups, and cosets of D_4.

Notice that in Figure 9 the left and right cosets do not coincide. In the special situation when left and right cosets are the same, the underlying subgroup H is called a *normal* subgroup. When the number of elements of a subgroup H is half that of the parent group G, there cannot be a difference between the left and right cosets. The subgroup H will be normal because the only cosets are $e \star H = H \star e$ and $G \setminus H = a \star H = H \star a$ for some element $a \notin H$. In this case there are two cosets, and so the subgroup has index 2. (In general, a subgroup that has exactly m cosets has index m.)

Figure 10 is a multicolor visualization of the elements, subgroups, and cosets of D_4. The upper-left and lower-right rosettes depict D_4 itself. The remainder of the left and right borders depicts individual elements of D_4, and the interior of the top and bottom rows depicts nontrivial subgroups of D_4. The multicolored rosettes in the second row show the left cosets corresponding to the subgroups in the top row; the multicolored rosettes in the third row show the analogous right cosets. The multicolored rosettes in the fourth row depict the cosets for the normal subgroups in the bottom row. (The reader is encouraged to verify this.) In fact, the different cosets, either left or right, are of the same symmetry pattern type as the original subgroup. In other words, if you look at just a single coset, a single color, it has the same symmetry pattern type as the original subgroup.

2.6 How Groups Act to Produce Two-Color Patterns

Consider a checkerboard pattern in black and white (see Figure 11(a)). There is no obvious background—are the black squares the pattern on a white background, or are the white squares the pattern on a black background? The entire region is decorated with two colors. Alternatively, we might have squares in two colors on a white background (see Figure 11(b)).

The salient feature of a two-color pattern is that every symmetry transformation either preserves the colors or reverses the colors. Symmetry transformations that either preserve or reverse colors are said to be *consistent with* color.

Not all patterns with exactly two colors appearing are true two-color patterns. For example, black polka dots on a blue background form a one-color pattern because there is no transformation that interchanges the black regions with the blue region. Thus, a *two-color pattern* is one in which all symmetry transformations are consistent with color and at least one symmetry transformation reverses color; a *one-color pattern* is one in

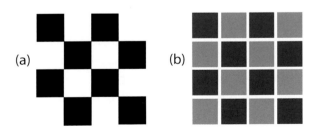

(a) (b)

Figure 11. Two types of two-color symmetry pattern using a checkerboard.

which all symmetry transformations are consistent with color and no transformation reverses color. With these definitions, Section 2.3 has secretly been analyzing group actions on one-color patterns. We now proceed to analyze group actions on two-color patterns.

In this case each group element acts either to preserve or reverse colors, which splits the group into two subsets—one is a color-preserving subgroup and the other is its color-reversing coset. Because there are two colors, the color-preserving subgroup is of index 2 (and thus normal).

Of course, there are sometimes different ways to assign two colors to a pattern. These correspond to different choices of index-2 subgroups. Therefore we notate a two-color symmetry pattern using both the underlying group G of the pattern and the chosen index-2 subgroup H. We write G/H, which experienced mathematical readers will recognize as the notation for quotient or factor groups. Each element of G/H represents a coset of H, so it has two elements (the identity element that preserves color and the transformation that reverses color) and is isomorphic to $\mathbb{Z}_2$.

When a pattern has m colors, we first check to see that for any two colors, there is a transformation that sends one to the other. This means there is an underlying transitive group action on the cosets formed by the m colors of the pattern. That is, the symmetry group of the pattern has transformations as elements. The subgroup corresponding to the color-preserving transformations is a normal subgroup, N; the index of the subgroup is m, and each element of the corresponding quotient group, G/N, corresponds to a different color in the pattern. For more on m-color patterns, see [5, Chapter 8] and [6].

For the remainder of this chapter, we will focus on two-color patterns. To find all two-color patterns, it suffices to search through the subgroups for each symmetry pattern to find all its index-2 subgroups. Recall that for one-color patterns, the fundamental domain differs from the translation unit when the pattern has reflection symmetry. Likewise, for two-color patterns one must be careful to examine the translation unit as well as the fundamental domain when seeking subgroups. This issue is exacerbated when there are two or more colors, as the translation unit may have to be enlarged. See Section 2.1 of Chapter 5 for details.

Additional information on two-color patterns can be found in [3], [5], [8], and [9]. Flow charts to help distinguish two-color patterns can be found in [8, pp. 129–162]. We continue our discussion of two-color patterns by examining two-color rosettes, frieze patterns, and wallpaper patterns.

2.7   Two-Color Rosette Patterns

Rosette patterns have only reflections and rotations (see [7]). If there is a single reflection line, as in Figure 12(a), the reflection acts to change colors across the reflection line, as in Figure 12(b).

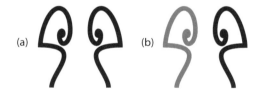

Figure 12. (a) Reflection symmetry and (b) two-color reflection symmetry.

If there are exactly two reflection lines, they must be perpendicular. Their actions produce three possible two-color patterns, although two of these are the same up to rotation. (See Figure 13.)

If there are four reflection lines, then they intersect at a central point at 45° angles. Again, their actions produce three possible two-colorings of this pattern, shown in Figure 14, and up to rotation two are the same.

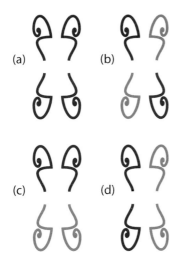

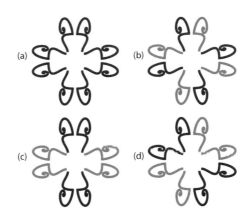

Figure 13. A pattern with two reflection lines: (a) shows the one-colored version and (b)–(d) show the three corresponding two-color patterns. Notice that (c) and (d) are the same up to rotation. Also notice that the purple portions of the figures in (b)–(d) are order-2 subgroups of (a).

Figure 14. A pattern with four reflection lines: (a) shows the single-colored version and (b)–(d) show the three corresponding two-color patterns. Notice that (c) and (d) are the same mathematically up to rotation.

One might think that Figure 15 shows a two-color pattern, but reflection across a diagonal does not reverse all the colors.

Figure 15. This pattern has four reflection symmetries but is not a true two-color pattern.

Because cross-stitch is done on a square grid as crosses ($\times$), a rosette pattern can have no more than four reflection lines and rotations no smaller than 90° [7]. A rosette without reflections has only rotational symmetries; the two-color patterns arising from the actions of rotation by 180° and 90° are shown in Figures 16 and 17, respectively.

This completes the classification of counted cross-stitch rosette patterns. Notice that Figure 10 contains some two-color rosette patterns, some four-color rosette patterns, and some rosettes that are not m-color patterns for any m.

Figure 16. A pattern with (a) a two-fold center of rotation and (b) its associated two-color pattern.

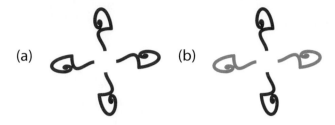

Figure 17. A pattern with (a) a four-fold center of rotation and (b) its associated two-color pattern.

2.8 Two-Color Frieze Patterns

There are seven frieze patterns [1,7]. We begin by acting on frieze patterns with translation or glide reflection to produce two-color patterns. The only way to two-color the translation-only frieze pattern is to alternate colors, as in Figure 18.

Figure 18. (a) Translation symmetry and (b) the two-color symmetry case.

Similarly, the only possible way to two-color the frieze pattern with just translations and glide reflections is to alternate colors as in Figure 19.

From these examples, one might expect that the number of two-color frieze patterns would be close to seven. But it turns out that there are seventeen two-color frieze patterns; when we include the seven one-color patterns (which are consistent with color, as no transformation reverses colors) we obtain a total of 24 one- and two-color possibilities. These are listed with examples in Table 2.

Figure 19. (a) Glide reflection symmetry and (b) the two-color symmetry case.

Two different notations for the two-color frieze patterns are given. The first notation originated with the crystallography community [5, Table 8.2.2]. It modifies the basic frieze pattern notation given in [7] by adding an apostrophe (′) to a symbol if the symbol corresponding to that transformation reverses colors, with one augmentation. When the third symbol is an *m*, and there is a glide reflection that preserves color, but the horizontal reflection denoted by *m* reverses color, then this *m* is changed to a *g*. The second notation (introduced in [4]) reflects our group action theme, and consists of two sets of symbols separated by a slash (/); the first gives the symmetry group of the underlying one-color pattern, and the second is the one-color pattern for the color-preserving index-2 subgroup.

Basic pattern	Two-color pattern (')	Two-color pattern (/)	Example
$p111$			
	$p'111$	$p111/p111$	
$p1g1$			
	$p1g'1$	$p1g1/p111$	
$p112$			
	$p112'$	$p112/p111$	
	$p'112$	$p112/p112$	
$pm11$			
	$p'm11$	$pm11/pm11$	
	$pm'11$	$pm11/p111$	
$p1m1$			
	$p'1m1$	$p1m1/p1m1$	
	$p1m'1$	$p1m1/p111$	
	$p'1g1$	$p1m1/p1g1$	
$pmg2$			
	$pm'g2'$	$pmg2/p1g1$	
	$pm'g'2$	$pmg2/p112$	
	$pmg'2'$	$pmg2/pm11$	
$pmm2$			

continued on next page

continued from previous page

Basic pattern	Two-color pattern (')	Two-color pattern (/)	Example
	$p'mm2$	$pmm2/pmm2$	
	$pm'm2'$	$pmm2/p1m1$	
	$pmm'2'$	$pmm2/pm11$	
	$pm'm'2$	$pmm2/p112$	
	$p'mg2$	$pmm2/pmg2$	

Table 2. The possible two-color frieze patterns along with their designations in two different systems. Some are demonstrated in Section 4.

2.9 Two-Color Wallpaper Patterns

In our analysis of two-color wallpaper patterns, we will generate examples by acting on a white square with a colored quarter circle in one corner (see Figure 20).

Figure 20. The 8 square $\times$ 8 square motif used in Figures 21–24.

Color-reversing transformations exchange the quarter circle with the background.

As with the frieze patterns, there are many more two-color patterns than one-color patterns. There are 17 one-color wallpaper patterns, of which only 12 can be stitched on counted cross-stitch fabric [7]. There are 46 two-color patterns [5, 8, 9]. Six of these arise from one-color wallpaper patterns that are not stitchable in counted cross-stitch because they have threefold or sixfold centers of rotation. The remaining 40 two-color patterns can be cross-stitched and are shown in Figures 21–24.

Like colors within each figure collect those two-color patterns that represent different quotients of the same G; that is, they are organized by the one-color patterns of which they are G-quotients (sometimes trivially).

In the $p4m$, pmm, $p4$, and $p4g$ patterns, we use fundamental domains that are modified from the motif given in Figure 20 in order to exhibit all symmetries in the quotient patterns.

Three notations for two-color wallpaper patterns are given here: the historical crystallography notation uses (') [8, pp. 70–71], and the slash notations both indicate quotient groups, one using the p, c, m, g symbols as before and the other using orbifold symbology [3]. All three generalize the notations for frieze patterns. Table 3 matches different notations for each group/subgroup with the corresponding image location.

Figure 21. Demonstration of *p*1-quotient two-color patterns in reds, *cm*-quotient two-color patterns in blues, *p*2-quotient two-color patterns in yellows, and *pgg*-quotient two-color patterns in greens.

Figure 22. Demonstration of *pmg*-quotient two-color patterns in blues and *pm*-quotient two-color patterns in reds.

Figure 23. Demonstration of *p4m*-quotient two-color patterns in purples and *cmm*-quotient two-color patterns in greens.

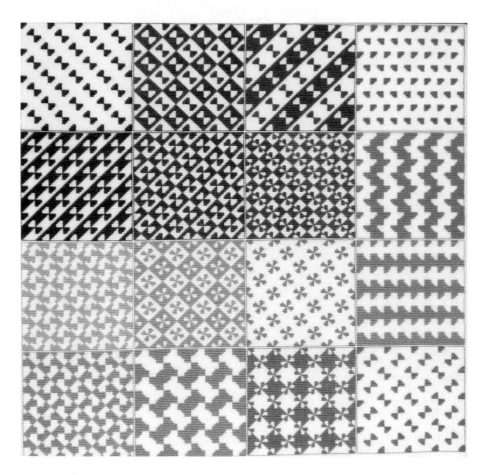

Figure 24. Demonstration of *pmm*-quotient two-color patterns in magentas, *pg*-quotient two-color patterns in oranges, *p4*-quotient two-color patterns in aquas, and *p4g*-quotient two-color patterns in greens.

Under-lying group	Notation with /	Conway notation	Historical notation	Figure and position of example by row and column
$p1$		○		Figure 21, R1 C1
	$p1/p1$	○/○	p'_b1	Figure 21, R1 C2
$p2$		2222		Figure 21, R3 C3
	$p2/p2$	2222/2222	p'_b2	Figure 21, R2 C3
	$p2/p1$	2222/○	$p2'$	Figure 21, R1 C3
cm		$*\times$		Figure 21, R2 C1
	cm/pm	$*\times\,/*\,*$	p'_cm	Figure 21, R2 C2
	cm/pg	$*\times\,/\times\times$	p'_cg	Figure 21, R3 C1
	$cm/p1$	$*\times\,/○$	cm'	Figure 21, R3 C2
pgg		$22\times$		Figure 21, R4 C3
	pgg/pg	$22\times\,/\times\times$	pgg'	Figure 21, R4 C1
	$pgg/p2$	$22\times\,/2222$	$pg'g'$	Figure 21, R4 C2
pmg		$22*$		Figure 22, R1 C1
	pmg/pmg	$22*\,/22*$	p'_bmg	Figure 22, R1 C2
	pmg/pm	$22*\,/*\,*$	pmg'	Figure 22, R1 C3
	$pmg/p2$	$22*\,/2222$	$pm'g'$	Figure 22, R2 C1
	pmg/pg	$22*\,/\times\times$	$pm'g$	Figure 22, R2 C2
	pmg/pgg	$22*\,/22\times$	p'_bgg	Figure 22, R2 C3
pm		$*\,*$		Figure 22, R4 C3
	$pm/pm(1)$	$*\,*\,/*\,*(1)$	p'_bm	Figure 22, R4 C2
	$pm/pm(2)$	$*\,*\,/*\,*(2)$	p'_b1m	Figure 22, R4 C1
	pm/cm	$*\,*\,/*\times$	$c'm$	Figure 22, R3 C3
	$pm/p1$	$*\,*\,/○$	pm'	Figure 22, R3 C2
	pm/pg	$*\,*\,/\times\times$	p'_bg	Figure 22, R3 C1
$p4m$		$*442$		Figure 23, R1 C1
	$p4m/p4m$	$*442/*442$	p'_c4mm	Figure 23, R1 C2
	$p4m/cmm$	$*442/2*22$	$p4'm'm$	Figure 23, R1 C3
	$p4m/pmm$	$*442/*2222$	$p4'mm'$	Figure 23, R2 C1
	$p4m/p4$	$*442/442$	$p4m'm'$	Figure 23, R2 C2
	$p4m/p4g$	$*442/4*2$	p'_c4gm	Figure 23, R2 C3
cmm		$2*22$		Figure 23, R4 C3
	cmm/pmm	$2*22/*2222$	p'_cmm	Figure 23, R4 C2
	cmm/pmg	$2*22/22*$	p'_cmg	Figure 23, R4 C1
	cmm/cm	$2*22/*\times$	cmm'	Figure 23, R3 C3
	$cmm/p2$	$2*22/2222$	$cm'm'$	Figure 23, R3 C2
	cmm/pgg	$2*22/22\times$	p'_cgg	Figure 23, R3 C1
pg		$\times\times$		Figure 24, R1 C4
	pg/pg	$\times\times\,/\times\times$	p'_b1g	Figure 24, R2 C4
	$pg/p1$	$\times\times\,/○$	pg'	Figure 24, R3 C4
pmm		$*2222$		Figure 24, R1 C1

continued on next page

continued from previous page

Underlying group	Notation with /	Conway notation	Historical notation	Figure and position of example by row and column
	pmm/cmm	$*2222/2*22$	$c'mm$	Figure 24, R1 C2
	pmm/pmm	$*2222/*2222$	$p'_b mm$	Figure 24, R1 C3
	pmm/pm	$*2222/**$	pmm'	Figure 24, R2 C1
	pmm/pmg	$*2222/22*$	$p'_b gm$	Figure 24, R2 C2
	$pmm/p2$	$*2222/2222$	$pm'm'$	Figure 24, R2 C3
$p4$		442		Figure 24, R3 C3
	$p4/p4$	$442/442$	$p'_c 4$	Figure 24, R3 C2
	$p4/p2$	$442/2222$	$p4'$	Figure 24, R3 C1
$p4g$		$4*2$		Figure 24, R4 C4
	$p4g/p4$	$4*2/442$	$p4g'm'$	Figure 24, R4 C3
	$p4g/cmm$	$4*2/2*22$	$p4'g'm$	Figure 24, R4 C2
	$p4g/pgg$	$4*2/22\times$	$p4'gm'$	Figure 24, R4 C1

Table 3. The possible two-color wallpaper patterns and their instantiations in cross-stitch.

3   Teaching Ideas

The original inspiration for the motif used in all of the two-colored wallpaper designs came from Mary's mother-in-law's quilt, shown in Figure 25. The basic block is called Drunkard's Path; note the quarter-circle cut out, just as in Figure 20.

Figure 25. The quilt that inspired a motif for two-color designs.

Although certain pairs of individual blocks show color reversal, the region shown of the quilt displays the one-color symmetry pattern $p4$. This suggests a one- and two-color symmetry pattern scavenger hunt. Participants should look for patterns in wallpapers, carpets, tiled floors, and quilts, in addition to other decorations. Public and private buildings are both excellent sources. The results can be recorded with cameras (use cell phones!). The location and supervision level can be modified to work for elementary through graduate classes.

Students of high-school age or older can find or create a motif and use it to design two-color patterns. Various media can be used for such designs. Colored designs on graph paper are comparatively quick to make and have similar design restrictions to cross-stitch. Indeed, the results can be used as cross-stitch patterns. Weavers might try instantiating two-color patterns in which only two colors are used (no background). Using techniques such as doubleweave or summer-and-winter will assure that the finished piece will be reversible. Tilers could use simple shapes such as squares or 45°-45°-90° triangles in different colors to cover walls or floors. For a less permanent tiling, cover a piece of paper or a bulletin board with colored paper tiles—a bulletin board decorated mathematically by students is an impressive show for parents or other visitors. Students can use any of the permanent techniques just mentioned to create a set of coasters with one- or two-color symmetry patterns. These can then be used by others to learn to identify the pattern types. Further, abstract algebra students can adopt one or more symmetry groups and explore two-colorings and associated subgroup structures. Students interested in additional challenge can investigate m-colorings and how they relate to subgroups.

In an abstract algebra class, students can create a visualization of the subgroups and cosets of D_4 with their own motif and colors, as was done in Figure 10. Shepherd has used this piece to assist students who are struggling with definitions and proofs involving subgroups and cosets. Together they examine the cross-stitched piece. Then Shepherd explains the representation of elements, the group operation, and the visualization of the first subgroup. The students usually take over at this point, and see that the other subgroups do satisfy the definition of subgroup. Next, Shepherd helps the students work towards an understanding of how cosets are represented. Together they determine which rosettes represent left cosets and which represent right cosets. This leads to conjecturing theorems, such as that a subgroup of index 2 is normal.

3.1   Exercises

Throughout the first set of exercises, students will explore the details of two-color patterns.

⋆ There are 52 wallpaper patterns shown in Figures 21–24. Verify that each pattern is of the symmetry type listed in Table 3. Note that some patterns are one-color patterns and others are two-color patterns.

Challenge students to determine the symmetry type of each pattern without referring to the figure captions or to Table 3. Students who know some group theory should examine the subgroups and cosets represented in each pattern.

⋆ In Figures 23 and 24, the fundamental domains for the *p4m*, *pmm*, *p4*, and *p4g* patterns, although related to the motif given in Figure 20, are not that motif. In each case, identify the fundamental domain for the pattern.

⋆ Figure 26 shows the project for this chapter. Each corner quadrant represents a different wallpaper pattern. The lower right panel shows the one-color pattern *p4g*. Determine the two-color pattern types of the other three corner quadrants.

⋆ The center panel of Figure 26 can be seen as a two-color pattern (reds form one color and greens form the other color). Identify this pattern. Considering only the red motifs also gives a two-color pattern. Which one? Similarly, focusing on the green motifs produces a two-color pattern. Identify this pattern as well. In addition, students of group theory may see that the center panel is a four-color pattern. Verify that this is the case and determine a transformation that functions as a group generator.

⋆ The borders between and surrounding quadrants and the center panel of Figure 26 form one- and two-color frieze patterns. Identify each one.

The next set of exercises will help abstract algebra students to visualize group-theoretic ideas.

⋆ Figure 8 depicts the subgroups of D_4. Identify each subgroup-of-another-subgroup in the figure. There are seven in all.

⋆ Find the center of D_4 in Figure 5.

Figure 26. A Two-Color Symmetry Sampler ready to be framed.

* Verify that in Figure 10 the multicolored rosettes in the fourth row depict the cosets for the normal subgroups of D_4.

* Any coset $a \star H$ or $H \star a$ is of the same symmetry pattern type as H. (See Figure 10 for examples.) Explain why this is the case.

3.2 Project Ideas

In this chapter, the group example used for visualization has been D_4 and the medium has been cross-stitch. What other finite groups can be visualized together with their subgroups and cosets using cross-stitch? How do the possibilities change if the medium is changed, for example to quilting? In particular, examine groups that are neither dihedral or cyclic. Can two-dimensional representations of three-dimensional symmetry groups (such as those for polyhedra) be realized in this way?

Start with a different one-color wallpaper pattern than used in the Section 4 sampler and generate all associated two-color wallpaper patterns. Incorporate these into a sampler, pillow, or quilt. For a smaller project, begin with a one-color frieze pattern, or cross-stitch all two-color rosette patterns. (We do not recommend stitching the entirety of a frieze pattern.)

4 Crafting the Two-Color Symmetry Sampler

The Two-Color Symmetry Sampler shown in Figure 26 is 128 stitches per side and covers a 9.14″ (23.22 cm) square of 14-count fabric.

Materials

* A 16″ square of 14-count Aida cloth, cream color. (You may need to buy a larger piece and cut it down to size.) To prevent the fabric from fraying, whip stitch around the edges or wrap them with narrow masking tape.

* A #24 tapestry needle.

* A 6″ embroidery hoop, to keep the fabric tight as one stitches.

* One skein of DMC floss in each of colors 434, 505, 742, 777, 931, 932, 989, 3328, and 3855. Use two of the six strands for stitching the cross-stitches and one strand for the backstitches. Work with 18″ lengths of floss. One way to organize unused floss is to insert floss into 2″ slits cut into a strip of cardboard, each slit labeled with a color name.

* (Optional, for pillow) $\frac{3}{8}$ yard of medium-weight backing fabric, such as a broadcloth.

* (Optional, for pillow) Polyester fiberfill or a 12″ pillow form.

Basic Cross-Stitch Instructions

Figure 27 shows the process of making cross-stitches. To begin, bring needle up at the start of the first stitch leaving a 1″ tail on the back side. Hold the tail in the direction you will be stitching, and cover it with the first few stitches to hold it in place. New threads can be started by passing the needle through several stitches on the back side of the embroidery. After you have worked an area, end your thread by weaving it through the wrong side of your stitches. Trim the loose thread ends neatly. (If you have ever seen the work of an experienced stitcher, the back is almost as neat as the front—there are no knots or loose thread tails.)

Backstitching (to outline the areas) is done after the cross-stitches are completed. To backstitch, anchor the thread on the back as for a new thread for cross-stitching. If you think of the holes in the fabric along the line where you are backstitching as numbered 1, 2, 3, etc., bring your needle to the top of the fabric through

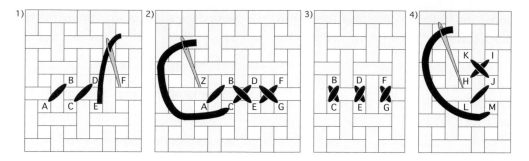

Figure 27. (1) Work all crosses through the holes of the fabric. Begin at the left and work to the right by coming up at A, going down at B, up at C, down at D, etc., across the row. (2) Complete the other half of each cross by stitching from right to left across the row, coming up at G, going down at D, coming up at E, etc., making sure all stitches cross in the same direction. (3) The back of the fabric is shown, although in reality the pairs of threads pull together to appear as single threads. (4) Work vertical stitches one at a time, following holes in alphabetical order H–L, then going down at J, up at M, down at H, etc.

hole 2, take the needle to the back through hole 1, bring the needle to the top through hole 3, take the needle to the back through hole 2, and continue in this way. End your thread as you did with a cross-stitch by weaving it through the wrong side of your stitches.

Symbol	DMC No.	Color Description
◢	742	Light Tangerine
✚	777	Wine Red
⧖	3328	Dark Salmon
▭	931	Blue Grey
◖	505	Pine Forest Green
#	3855	Autumn Gold
◖	932	Seagull Blue
∩	989	Fennel Green

Figure 28. The key for Figures 29–32.

Sampler Instructions

Figures 29–32 show the charts for the Two-Color Symmetry Sampler, with the center of the pattern marked with arrows and overlap between charts shaded in blue-grey. Figure 28 is the key. It is usually easiest to start the design at the center and work out to the edges. (To find the center, fold your fabric in half vertically, then horizontally. Where the folds meet is the center point.) Mark the center with a pin or removable piece of thread.

Use two strands of thread for regular cross-stitches. Use one strand of thread for backstitch of Pine Forest Green (DMC 505), Wine Red (DMC 777), Cigar Brown (DMC 434), and Blue Grey (DMC 931) for the appropriate colored lines in the pattern.

Cleaning and Pressing

Use a mild dishwashing liquid in lukewarm water to hand wash your design. Cold water washes made for fine washables and wool garments may actually cause the thread to bleed. Rinse in lukewarm water until the water is clear. Roll the design in a towel and squeeze well to remove excess water. Lay your design flat to air dry. To press your design after it has dried, lay it face down on a towel. Use a dry iron on a medium temperature setting and press lightly so you do not flatten your stitches.

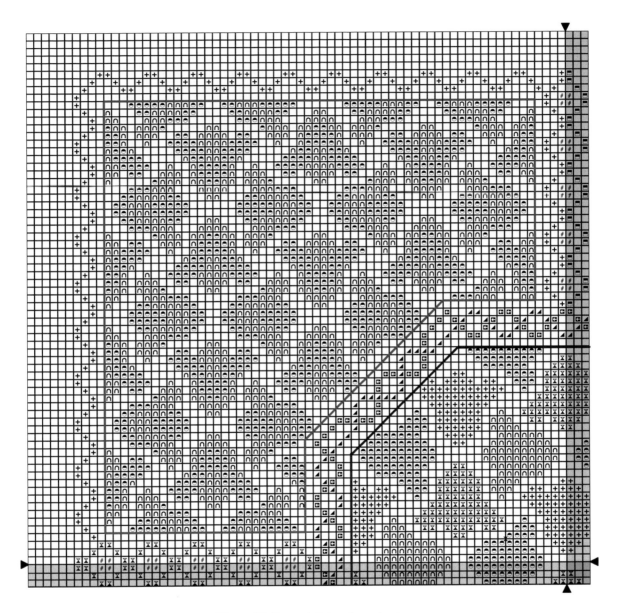

Figure 29. Upper-left-hand quadrant.

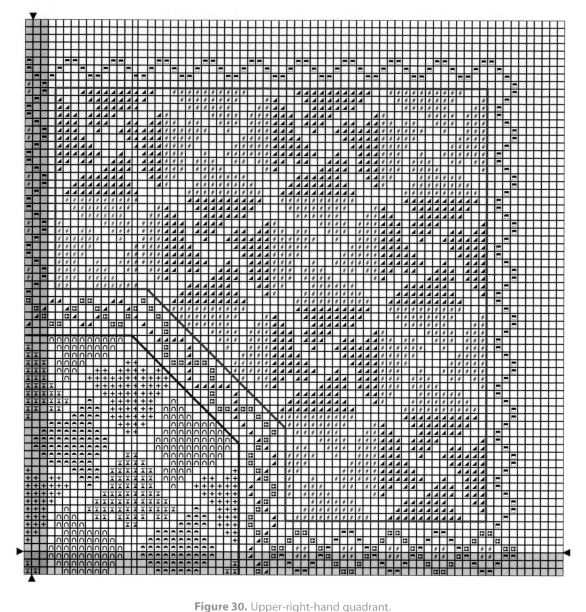

Figure 30. Upper-right-hand quadrant.

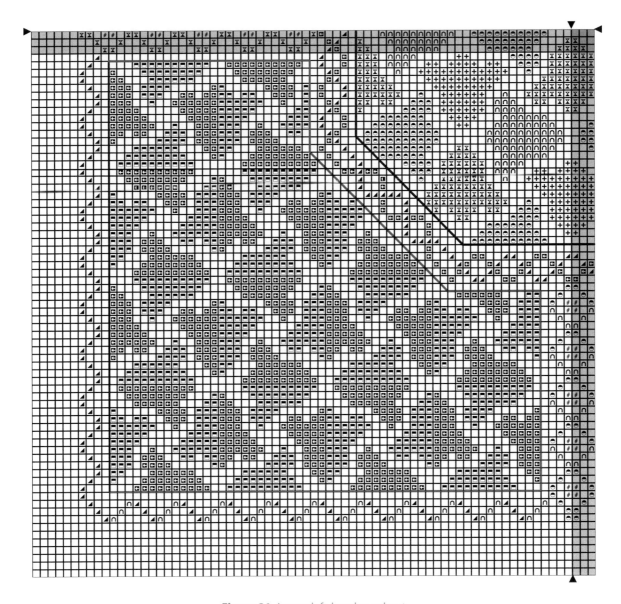

Figure 31. Lower-left-hand quadrant.

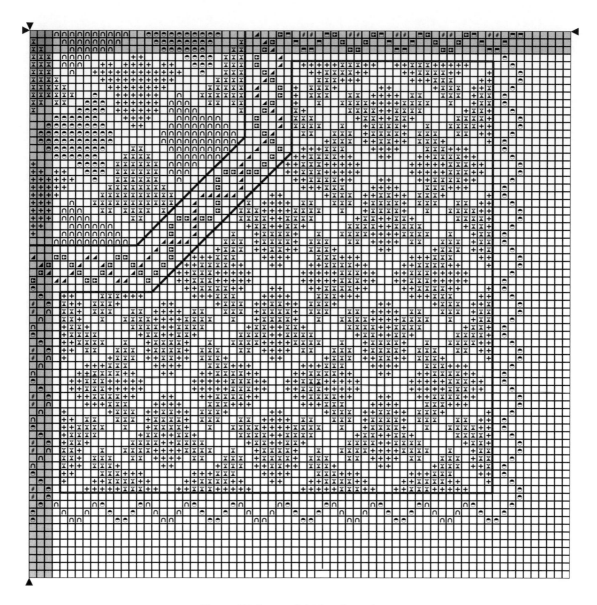

Figure 32. Lower-right-hand quadrant.

Finishing

The completed piece may be finished as a pillow or matted and framed. Consult a needlecraft or frame shop for professional framing. To complete as a pillow, trim the needlework and backing fabric to 13″ square. Sew the backing fabric to the needlework with a $\frac{1}{2}″$ seam allowance. Be sure to seam along the grain of the Aida fabric. Leave a 4″ opening at the bottom. Turn the pillow right-side-out and stuff with fiberfill or insert pillow form. Hand-sew opening closed.

Bibliography

[1] belcastro, sarah-marie and Hull, Thomas. "Classifying Frieze Patterns without Using Groups." *The College Mathematics Journal* 33:2 (2002), 93–98.

[2] Carter, Nathan. *Visual Group Theory*. The Mathematical Association of America, Washington, DC, 2009.

[3] Conway, John Horton, Burgiel, Heidi, and Goodman-Strauss, Chaim. *The Symmetries of Things*. A K Peters, Wellesley, MA, 2008.

[4] Coxeter, H. S. M. "Coloured Symmetry." In *M.C. Escher: Art and Science*, edited by H. S. M. Coxeter, Michele Emmer, Roger Penrose, and M.L. Teuber, pp. 15–33. North-Holland, Amsterdam, 1986.

[5] Grünbaum, Branko and Shephard, G. C. *Tilings and Patterns*. W. H. Freeman and Company, New York, 1987.

[6] Jarratt, J. D. and Schwarzenberger, R. L. E. "Coloured Plane Groups." *Acta Crystallographica* A36 (1980), 884–888.

[7] Shepherd, Mary. "Symmetry Patterns in Cross-Stitch." In *Making Mathematics with Needlework: Ten Papers and Ten Projects*, edited by sarah-marie belcastro and Carolyn Yackel, pp. 71–89. A K Peters, Wellesley, MA, 2008.

[8] Washburn, Dorothy and Crowe, Donald. *Symmetries of Culture*. University of Washington Press, Seattle, 1988.

[9] Woods, H. "The Geometrical Basis of Pattern Design," *Journal of the Textile Institute*, 27 (1936), 305–320.

CHAPTER 7

*perfectly simple:
squaring the rectangle*

SUSAN GOLDSTINE

1 The Math and Motivation behind the Pattern

One of the hazards of being a mathematician is that one never knows when one will be seized by a new idea that refuses to let go. It could happen while admiring a work of art, or strolling by a babbling brook, or passing a new gadget in a hardware store ... or stumbling across the latest crochet craze while surfing the web. What caught my eye this time was the Babette blanket [8], an eye-catching pattern of granny squares in different sizes and colors stitched together into a rectangular blanket.

When I first saw the blanket, it immediately entered the queue of things I should make—sometime. I love the idea of carefully selecting colors and mapping out where I want to put them, all the while giving in to the studied randomness of the design. And someday, when I have an awful lot of time, I will make one. But I was also gripped by a mathematical notion, one that could not wait.

In a Babette blanket, each square size is repeated many times. What about making a rectangle out of squares of different sizes with no repeats? Scholars of mathematics have studied the problem of dividing rectangles into squares of distinct sizes for many years and have devised many lovely examples [7], a few of which are pictured in Figures 1, 2, and 4. The result of crocheting one is Perfectly Simple, the project presented here.

There are spectacular renditions of knitted dissected rectangles, most notably a wall hanging in the London Science Museum from the pattern Square Deal [2]. (A commercial knitting pattern for a Square Deal cushion is available in [3].) However, crochet offers pattern opportunities that knitting does not, and each craft brings its own charms to the dissection of rectangles into squares.

1.1 Dissections of Rectangles

A rectangle that has been cut into finitely many squares is known as a *squared rectangle*. In the special case where the original rectangle is itself a square, the result is a *squared square*. In general, when dissecting a rectangle this way, there will be squares of duplicate sizes. However, by being clever, one can dissect certain rectangles into squares all of distinct sizes. Such a rectangle is a *perfect squared rectangle*. Some squared rectangles can be decomposed into unions of smaller squared rectangles; if no such decomposition exists, the rectangle is a *simple squared rectangle*.

Perfectly Simple depicts the smallest simple perfect squared rectangle. There are no perfect squared rectangles with fewer than nine squares and there are exactly two perfect squared rectangles with exactly nine squares [4]; of those two rectangles, Perfectly Simple has the smallest proportions, 32×33. The other, pictured in Figure 1(a), is 61×69. These results and many more are catalogued in [1].

Woolly Thoughts' knitted Square Deal depicts the smallest simple perfect squared square, as shown in Figure 2. Discovered by A. J. W. Duijvestjin [5], this squared square has side length 112 and is dissected into 21 smaller squares. A crocheted version of Duijvestjin's squared square would make a spectacular project for the patient crocheter with a lot of free time.

The rectangle in Perfectly Simple is divided into squares of side lengths 1, 4, 7, 8, 9, 10, 14, 15, and 18. One of the advantages of making this rectangle using crochet in the round is that the striping of the rounds shows the square sizes clearly. Each square has side length equal to the number of stripes, where each stripe is a half-unit thick.

A nice additional feature of Perfectly Simple is that it displays a three-coloring of the squared rectangle it

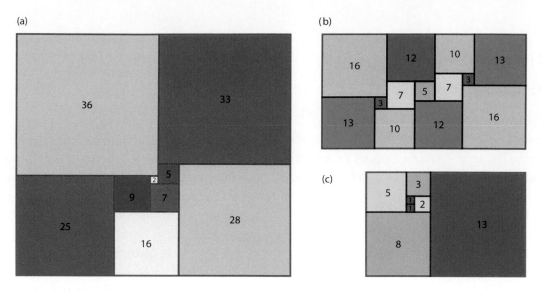

Figure 1. Squared rectangles: (a) perfect and simple, (b) imperfect and simple, and (c) neither perfect nor simple.

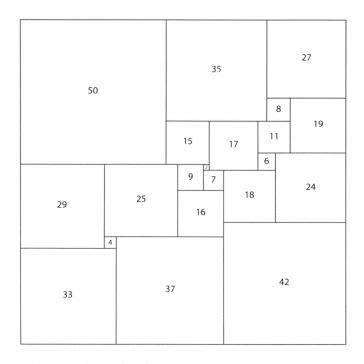

Figure 2. The smallest known simple perfect squared square.

represents. It is possible to color the nine squares with three different colors so that no two squares that share part of a border have the same color. This is not true of squared rectangles in general, and we invite the reader to verify that the Square Deal squared square requires at least four colors.

2 Crafting Perfectly Simple

Kathy Merrick's original Babette blanket [8] is a riot of squares of many different colors and sizes. The different square sizes create a pleasing sense of randomness, but while perusing the blanket one's eyes are continually drawn to clumps of squares of the same size. As a mathematician, I found it impossible not to make the leap to a blanket of squares in which no two squares are the same size. Since a blanket is a rectangle, the result is a simple perfect squared rectangle.

I wanted the arrangement and sizes of the squares to stand out, and so I opted for a minimal color scheme. Unlike the rainbow of Babette blankets festooning the internet, Perfectly Simple uses just enough colors to stripe the squares so the sizes are clearly visible and to make touching squares different colors.

Squares are a mainstay of traditional crochet, so making a crocheted squared rectangle is a fairly straightforward process: crochet solid granny squares of the appropriate sizes, stitch them together, and stabilize the edges by crocheting a border around the entire rectangle. The techniques in this section are easily adaptable to any squared rectangle, and Perfectly Simple can be crocheted with any yarn in any gauge.

2.1 Pattern Overview

Materials

Seven colors of yarn, in the following amounts.

* For a decorative version (pictured in Figure 3) measuring 21.5″ by 22″, use sport-weight yarn.

  - Colors 1 and 2, 200 yds.
  - Colors 3 and 4, 150 yds.
  - Colors 5 and 6, 100 yds.
  - Color 7, 20 yds.

Figure 3. Perfectly Simple.

- For a baby blanket measuring 37″ by 38″, use worsted-weight yarn and size I hook.

    - Color 1, 4.5 oz. (245 yds).
    - Color 2, 4 oz. (217 yds).
    - Color 3, 3.5 oz. (190 yds).
    - Color 4, 3 oz. (163 yds).
    - Color 5, 2.5 oz. (136 yds).
    - Color 6, 2 oz. (110 yds).
    - Color 7, 1.5 oz. (80 yds).

Abbreviations

- Sl st (or sl st) means slip stitch.
- Ch (or ch) means chain.

- Hdc (or hdc) means half double crochet.
- Dc (or dc) means double crochet.

Color Notes

- Perfectly Simple uses six yarn colors in three pairs, plus a seventh color for the edging. In Figures 3 and 4, the colors are as follows:

    - Color 1:   dark blue,
    - Color 2:   light blue,
    - Color 3:   dark red,
    - Color 4:   light red,
    - Color 5:   dark yellow,
    - Color 6:   light yellow,
    - Color 7:   white.

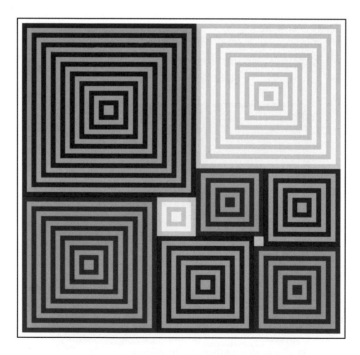

Figure 4. Diagram of Perfectly Simple.

Each square is crocheted in one pair of colors, with the rounds alternating between the two colors, and the final round of each square is always the darker color in the pair. As mentioned in the previous section, the size of each square in a crocheted squared rectangle is the same as the number of crochet rounds. Since the rounds alternate colors, the size is also the number of visible stripes in the square.

The nine squares in the pattern come in the following sizes and colors.

Size	Color Pair	Starting Color
1	(5, 6)	5
4	(5, 6)	6
7	(3, 4)	3
8	(1, 2)	2
9	(3, 4)	3
10	(1, 2)	2
14	(3, 4)	4
15	(5, 6)	5
18	(1, 2)	2

2.2   Granny Square Recipe

There are many eloquent patterns for solid granny squares readily available, and you may find it helpful to consult a complete crochet source. For instance, see the lovely video tutorial [9]. Here, we give a basic recipe for a solid granny square, along with some tips that are specific to this pattern.

Foundation Ring: With starting color, ch 4, sl st to close ring.

The inevitable hole in the center of crochet in the round throws off the proportions of the granny squares. For instance, a 1×1 granny square will be wider than one eighth of an 8×8 granny square, because while the 8×8 square has eight times as many rounds, the central gap in the 1×1 square is just as large as the gap in its larger counterpart. This creates a slight discrepancy in the assembly process, in which the 1×1 and 7×7 square have to be sewn to one edge of the 8×8 square. Therefore, it is worth making the foundation ring as small as you comfortably can.

Round 1: Still with starting color, ch 3 (counts as first dc), dc; * ch 2, dc 3; repeat from * twice; ch 2, dc, sl st to close round, and finish off.

All double crochets are stitched into the center of the foundation ring. After round 1, the square has three double crochets on each side and a chain 2 in each corner.

Subsequent Rounds: With new color, start with ch 3 (counts as first dc); work 1 dc into each dc of the previous round, and work dc 2, ch 2, dc 2 into each ch 2 in the previous round; sl st to close round, and finish off.

Notice that in each round, four double crochets are added to each side of the square. Thus, the total number of double crochets on each side of the round are as in Table 1; there will be a chain 2 at each corner between sides.

Since the yarn color changes with each new round, there is no easy way to get around weaving in a ton of ends. On the other hand, this allows you to change the starting position of each new round to distribute the

Round	1	2	3	4	5	6	7	8	9	10
dc's per side	3	7	11	15	19	23	27	31	35	39

Round	11	12	13	14	15	16	17	18	n
dc's per side	43	47	51	55	59	63	67	71	$4n - 1$

Table 1. Double crochets on each round.

yarn changes evenly around the square, which makes both the beginning-of-round chain stitches and the woven-in ends less conspicuous in the completed square.

Blocking

It is *extremely important* to block the squares before you proceed with the rest of the pattern. The sizes of the squares must be consistent, so you need to decide how large one unit is based on how big the squares are before blocking and carefully measure as you pin so that each $n \times n$ square is blocked to a side length of n units. Blocking can stretch crochet but not contract it; choose your unit length accordingly.

2.3   Rectangling the Squares

Once the squares are blocked to size, you are ready to stitch them together. Mattress stitch works best for a seam that is virtually invisible from both sides. If you are unfamiliar with mattress stitch, you may find the video tutorial [6] helpful; even though the demonstration is on knit pieces, the same method works for crochet. The tricky part of the process is the alignment of stitches. By and large, the stitches on either side of the seam match up one-to-one, but there is a complication at the corners.

Consider, for example, the edge of the 8×8 square that is stitched to the 7×7 and 1×1 squares, as shown in Figure 5. The edge of the 8×8 square has 31 double crochet stitches, while the edges of the smaller squares have 27 and 3 double crochet stitches, respectively, for a total of 30. This means that the corners of both of the smaller squares must be stitched to the same stitch in the 8×8, since there is only one double crochet left over.

Figure 5 shows the general method applied to all T-joins in a crocheted squared rectangle. It is best to stitch the stem of the T (in this case, the yellow-red edge) before the top. Where two squares meet each other at a corner, one chain from the first square is stitched to one chain from the second. Then the remaining chains are both attached to a single stitch on the longer edge they meet.

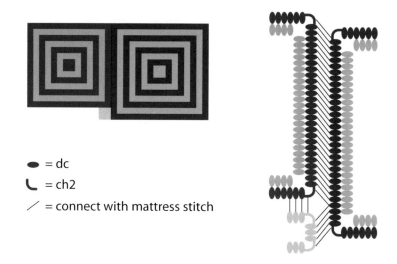

● = dc

∪ = ch2

╱ = connect with mattress stitch

Figure 5. Mattress stitching the squares together.

Mattress stitch is pretty well concealed, so in principle you can seam with whatever yarn you like. However, to gild the lily, you can make both the seam stitches and the woven-in ends completely invisible by sewing each seam with the yarn color of the square on one side of the seam and then weaving the ends into that square. There are a couple of seams that must be split into two sections for this method to work, but the final effect is worth the extra effort. Figure 6 shows which colors are used for which seams. At this point, blocking the rectangle to even out the edges is helpful but not essential.

Figure 7. A simpler Perfectly Simple.

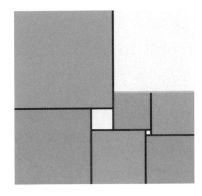

Figure 6. Hiding the seams.

Border

Using color 7, start in the middle of the edge of any square and ch 2. In each dc, work one hdc. At each place along the edge of the rectangle where two square corners meet, hdc 2 tog, working one stitch into each corner. At each corner of the rectangle, work hdc 1, dc 1, hdc1. When you return to your starting point, sl st to close the round and finish off.

2.4 Alternate Construction

The variant on Perfectly Simple shown in Figure 7, designed by Carolyn Yackel, uses traditional granny squares and fewer yarn colors.

Materials

★ Three colors of yarn, in the following amounts.

- Color 1, 150g/5.25 oz. (420 yds).
- Color 2, 100g/3.5 oz. (280 yds).
- Color 3, 100g/3.5 oz. (280 yds).

★ Size I crochet hook.

Traditional Granny Square Construction

All pattern notes follow those for the solid granny square construction, which should be read regardless of which pattern is being followed. Squares of sizes 10 and 18 are worked in color 1; squares of sizes 7, 9, and 14 are worked in color 2; and, squares of sizes 1, 4, 8, and 15 are worked in color 3.

Foundation Ring: Ch 4, sl st to close ring.

Round 1: Ch 3 (counts as first dc), dc; * ch 3, dc 3; repeat from * twice; ch 3 dc, sl st to close round.

Round 2: Sl st *to the right* into the corner space. Ch 3 (counts as first dc), dc; * ch 2, work (3 dc, ch 3, 3 dc) in next corner space; repeat from * twice; ch 2, work 3 dc, ch 3, dc, sl st into top of initial ch 3 of round to join and complete round.

square size	1	4	7	8	9	10	14	15	18
sc + ch on finished edge	6	24	42	48	54	60	84	90	108

Table 2. Squares plus chains per square size.

Subsequent Rounds: Sl st *to the right* into the corner space. Ch 3 (counts as first dc), dc. (This leaves the half corner incomplete—it will be completed at the end of the round.) † Ch 2, *work 3 dc in next ch 2 sp, ch 2. Repeat from * across the edge. Work (3 dc, ch 3, 3 dc) into corner space. Repeat from † twice. Ch 2, *work 3 dc in next ch 2 sp, ch 2. Repeat from * across the edge. When the final/beginning corner is reached, work 3 dc, ch 3, dc. Sl st into top of initial ch 3 to close round. When finished size is reached, make single crochet border without breaking the thread.

Single Crochet Border

Without chaining 1, single crochet into each dc, 3 sc into each ch 2 space, and sc, ch 1, sc into each corner space, sl st into first sc at end of round to join, cut thread, and weave in end. Notice that even though each end of each side has a chain to turn the corner, this chain is shared between two sides so only one chain is counted per side (see Table 2).

Join for Squares

This time-consuming method looks like a zig-zag. The result is pretty, but does not disappear, so changing attaching yarn color to match one of the adjacent square colors is recommended.

Place the squares right side up and next to each other in correct position, referring to the diagram in Figure 4. Now make a slip knot in the yarn and put on the hook. Slip stitch through the back loops of the squares to be joined *one at a time*. First the left, then the right, then the left, then the right. Only go through one loop at a time. Try to keep an even tension on the slip stitches. When it is time to change colors, cut the yarn, pull through, and weave in the end.

Bibliography

[1] Anderson, Stuart. *Tiling by Squares*. Available at http://www.squaring.net/sq/tws.html, 2010.

[2] Ashforth, Pat and Plummer, Steve. "Square Deal." *Woolly Thoughts Afghans*. Available at http://www.woollythoughts.com/afghans/deal.html, accessed October 30, 2010.

[3] Ashforth, Pat and Plummer, Steve. *12 Pillows of Wisdom*. Assign Publications, Colne, Lancashire, UK, 2001.

[4] Brooks, R. L., Smith, C. A. B., Stone, A. H., and Tutte, W. T.. "The Dissection of Rectangles into Squares." *Duke Mathematical Journal* 7 (1940), 312–40.

[5] Duijvestijn, A. J. W. "A Simple Perfect Square of Lowest Order." *J. Combin. Theory Ser. B* 25 (1978), 240–243.

[6] Findlay, Amy. "Knitting Tips." *KnittingHelp.com* Available at http://www.knittinghelp.com/videos/knitting-tips, 2010.

[7] Gardner, Martin. "Squaring the Square." In *The Second Scientific American Book of Mathematical Puzzles and Diversions*, pp. 186–209. University of Chicago Press, Chicago, 1987.

[8] Merrick, Kathy. "Babette Blanket." *Interweave Crochet* (Spring 2006), 22. (Also Available at http://www.interweavestore.com/Crochet/Patterns/Babette-Blanket.html.)

[9] Richardson, Teresa. *Crochet Solid Granny Square*. Available at http://www.youtube.com/watch?v=gJCd_-V2D5k, 2007.

CHAPTER 8

spherical symmetries of temari

CAROLYN YACKEL

WITH SARAH-MARIE BELCASTRO

1 Overview

The ancient craft of temari comes from Japan, whence it evolved over centuries from mothers making toy balls for children to artists desiring to precisely decorate ornaments. A particularly fine history is contained in [14]. Temari balls are spherical forms wrapped in thread, upon which designs are then wrapped or embroidered in perle cotton to create striking patterns. Sometimes these patterns are representational, but we will focus on the many patterns that divide the sphere into geometric shapes. Yackel has studied temari for years, and the work presented here builds on her earlier publications [16, 17]; belcastro contributed content only to Section 2.

The majority of temari designs are highly symmetric: if a set of geometric shapes appears once, it repeats all over the surface, covering the sphere completely and leaving no gaps. This is called a *tiling* of the sphere. The word *tiling* here is used much like it is for a bathroom or kitchen floor. For such purposes we usually use flat tiles, but these wouldn't conform to the surface of a sphere. On the other hand, if we had a tiling on a sphere made of soft clay, we could then flatten each tile (perhaps by smacking it with a pan) and make our previously tiled sphere have a surface with flat sides, as depicted in Figure 1. If we then also straightened out the edges, the resulting object would be called a polyhedron. Colloquially, a *polyhedron* is a many-sided three-dimensional object, where each face is flat and has straight edges. Conversely, we could inflate a polyhedron until it is projected onto a sphere. For example, imagine (see Figure 2) constructing the skeleton of an octahedron using flexible edges and putty corners. Now imagine putting a balloon in the middle of the octahedron and inflating the balloon until it forms a sphere with the putty corners just touching the surface of the balloon. Indeed, these edges now form lines on the sphere, connecting the points corresponding to the putty corners. The transformation in Figure 2 describes the radial projection of an octahedron onto a sphere. The symmetries of the projected octahedron are the same as the symmetries of the octahedron.

Therefore, we may consider certain tilings on the sphere to be equivalent to tilings on polyhedra.

Of greatest interest to temari artists are the most symmetric spherical tilings. These are *regular*, meaning that every tile is the same regular polygon and the

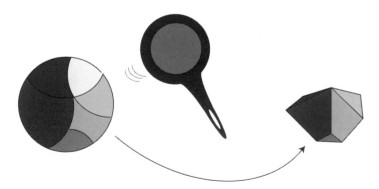

Figure 1. A tiling on a sphere can be flattened out into a polyhedron using a frying pan or some other appropriate mapping. The yellow face is no longer visible.

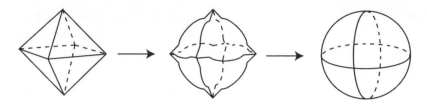

Figure 2. An octahedron morphs via radial projection to a sphere.

structure near each vertex is identical. Mathematicians will recognize the corresponding polyhedra as the Platonic solids pictured in Figure 3.

Some counting of polyhedral pieces in Figure 3 will reveal interesting facts:

* The number of vertices of the cube (eight) is the same as the number of faces of the octahedron, and the number of vertices of the octahedron (six) is the same as the number of faces of the cube.

* The number of edges of the cube (12) is the same as the number of edges of the octahedron.

* The number of vertices of the icosahedron (12) is the same as the number of faces of the dodecahedron, and the number of vertices of the dodecahedron (20) is the same as the number of faces of the icosahedron.

* The number of edges of the dodecahedron (30) is the same as the number of edges of the icosahedron.

* The number of vertices of the tetrahedron (four) is the same as the number of faces of the tetrahedron. (Of course, the number of edges of the tetrahedron (six) is the same as the number of edges of the tetrahedron.)

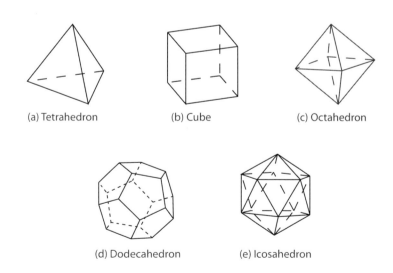

(a) Tetrahedron (b) Cube (c) Octahedron

(d) Dodecahedron (e) Icosahedron

Figure 3. The Platonic solids.

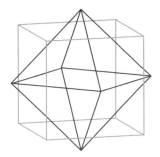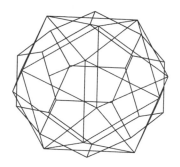

Figure 4. The Platonic solids pictured as dual pairs.

That these numbers match up nicely is a consequence of the polyhedral pairing known as *duality*. When two polyhedra are dual, the vertices of one polyhedron correspond to the faces of the other (and vice versa) and the edges of one polyhedron correspond to the edges of the other. In Figure 4, this pairing is shown for the Platonic solids. Notice that the vertices of the octahedron appear to poke through the centers of the faces of the cube and vice versa (left), that the vertices of the icosahedron appear to poke through the centers of the faces of the dodecahedron and vice versa (middle), and that the vertices of one tetrahedron seem to poke through the centers of the faces of the other (right). Corresponding edges appear to cross each other at right angles.

An amazing fact of temari balls is that when one of these basic polyhedra is projected onto a ball, its dual typically appears as well. This is explained below in Section 2.1, where duality is defined for spherical polyhedra as mathematical stand-ins for temari balls.

An artist creates a temari ball by wrapping a base sphere with thread, then wrapping this with carefully placed contrasting *guideline* threads that determine the desired spherical tiling, and finally employing these contrasting threads as guides for embroidering motifs into the wrapping thread. A mathematician might begin by situating an ideal sphere in space and determining

the coordinates of the vertices of the polyhedron she wanted to project onto its surface. From this perspective, it seems that one only needs to place pins at these coordinate points to make guidelines for the desired polyhedron. Unfortunately, the physical imperfection of thread-wrapped spheres creates a gap between the mathematical idealization of the scenario and reality. Furthermore, as the world we live in is not coordinatized, the ball does not inhabit a specific coordinate framework.

A natural set of coordinates would be given with respect to the radius and use irrational numbers. Without cutting open the wrapped sphere, which would ruin the wrapping, the artist cannot determine its radius. Furthermore, it is not possible to accurately measure irrational numbers. The ball's circumference is necessarily the only available measurement. Therefore, plotting points on the surface of a temari ball is mathematically nontrivial. Yet, exact point placement is of paramount importance as misplacement will result in a lack of symmetry in guideline wrapping and will ruin the beauty of the ball. The mathematics of point placement is discussed in Sections 2.3 and 2.4.

Some finite spherical symmetry types are much more common in temari than others. Nonetheless, all of the 14 finite spherical symmetry types [2] are achievable in temari, as is illustrated in Section 2.5. In Section 2.6 we discuss a temari-specific refinement of the finite spherical symmetry classification system. Section 3 contains directions for teachers wanting to make temari balls with their classes. Questions are included to direct students' mathematical thinking throughout the process. In Section 4 instructions are given for three different temari balls. The first two are appropriate for the classroom and are intended to be used in conjunction with the teaching section, and the third is an example of the ideas developed in Section 2.4.

2 Mathematics

The designs of temari balls are embellished spherical tilings, where the base tilings are derived from underlying polyhedra projected to the surface of the sphere. Technically a *polyhedron* is a compact convex three-dimensional body with a finite number of extremal points (called *vertices*). A *spherical polyhedron* is the projection of the edges of a polyhedron to the surface of a sphere. For the purposes of this chapter, we will only examine polyhedra that are vertex-transitive (or face-transitive), meaning that for any two vertices (respectively faces) there exists a symmetry transformation of the polyhedron that takes one vertex (respectively

face) to the other. This condition assures that a temari ball based on such a polyhedron will meet the desired symmetry aesthetic. A vertex-transitive polyhedron has all vertices lying on a sphere; the associated spherical polyhedron is also vertex-transitive. Notice that the Platonic solids are both vertex- and face-transitive, the Archimedean solids are only vertex-transitive, and the Catalan solids are only face-transitive. These three sets of solids will be our primary examples.

2.1 Duality Exhibited by Temari Balls

The theme of duality flows across areas of mathematics, and so there are many types of duality. In the case of vertex- or face-transitive polyhedra the appropriate definition is reciprocation with respect to the circumscribing sphere (equivalently polar duality) [6]. Polar duality is demonstrated for the Platonic solids in Figure 4. Conveniently, when we project a polyhedron onto the sphere, the analogue of polar duality can be described simply. The *spherical dual* of a spherical polyhedron is formed by placing a vertex in the center of each face and connecting adjacent faces with arcs of great circles. This is shown for dual pairs of spherical Platonic solids in Figure 5. (The dual polyhedron associated to this spherical dual may be obtained by taking the convex hull of the spherical dual's vertices.)

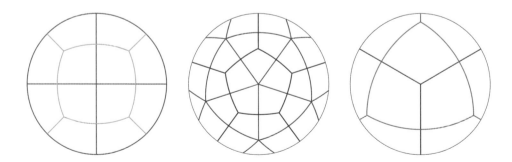

Figure 5. The spherical Platonic solids pictured as dual pairs to match Figure 4.

An *incircle subdivision* of a polyhedron is formed by bisecting every facial interior angle and dropping perpendiculars from the resulting center to every edge of each polyhedral face. See Figure 6 for an illustration on one triangular face.

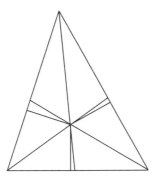

Figure 6. An incenter found by intersecting black angle bisectors has edge perpendiculars in red.

Such a subdivision only exists when every face of the polyhedron has an incircle, that is, when the interior angle bisectors meet at a single point on each face. This is the case for faces that are regular polygons. Projecting the incircle subdivision to the circumscribing sphere and extending the arcs to great circles produces *incircle guidelines*, as shown in Figure 7. These definitions arose while developing the following result.

Theorem 1 *The incircle guidelines of a spherical polyhedron with regular-polygon faces produce a set of guidelines for the dual spherical polyhedron.*

Proof Focus on a single face of a spherical polyhedron each of whose faces are regular polygons. (For visual aids, see Figures 5 and 7.) The arcs resulting from the interior angle bisectors of this face intersect in the center of the face, which marks a vertex of the dual spherical polyhedron. Adjacent faces on the spherical polyhedron share an edge. Because these faces are regular polygons, the perpendiculars dropped from the centers of the two faces bisect this edge and join together to form a single edge of the dual spherical polyhedron. (Note that for a regular polygon, the incircle subdivision coincides with the barycentric subdivision.) Thus, an arc subset of the incircle guidelines forms the edge set of the dual spherical polyhedron.   □

One visual consequence of Theorem 1 is that a ball exhibiting a cube tiling also naturally has octahedral symmetry. At left in Figure 8, we see a spherical cube formed by the silver vertices with a green-outlined cross embellishing the center of each face; rotating the ball places a silver star motif in the center of each octahedral face, with cross centers marking vertices of a spherical octahedron. This change in perspective refocuses our attention from the square regions outlining the

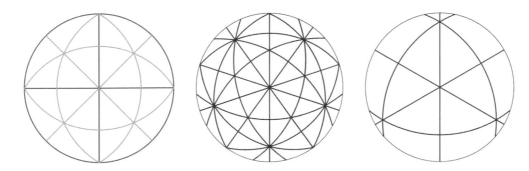

Figure 7. Incircle guidelines resulting from the spherical Platonic solids in Figure 5.

Figure 8. Cube (left) and octahedral (right) symmetry exhibited on the same temari ball.

projected cube to the triangular regions outlining the projected octahedron. Examples of other balls exhibiting spherical Platonic dual pairs are given in Figure 9.

Corollary 1 *If a polyhedron has a dual with regular-polygon faces, then the incircle guidelines for the spherical polyhedron produce guidelines for the dual spherical polyhedron.*

Proof By Theorem 1, the incircle guidelines of the dual spherical polyhedron are guidelines for the original spherical polyhedron. Thus, the vertices of the dual spherical polyhedron are marked as intersections of interior-angle bisectors from faces on the original spherical polyhedron. Moreover, the guidelines connecting two adjacent faces of the original polyhedron are perpendicular to their common edge, which shows them to be segments of an incircle subdivision of the original (spherical) polyhedron. Therefore the dual spherical polyhedron guidelines are incircle guidelines of the original polyhedron. □

Figure 9. A tetrahedral self-dual ball (left), a cube/octahedral ball (center), and a dodecahedral/icosahedral ball (right).

Students of polyhedra will now recognize that because the presence of dual incircle guidelines requires that a polyhedron or its dual must have regular faces, and because we are considering only vertex- and face-transitive polyhedra, our study is restricted to the Platonic solids, the Archimedean polyhedra, and the Catalan solids.

It follows from Corollary 1 that a completed set of incircle guidelines may be viewed either from the perspective of marking a specific polyhedron or of marking its dual. (Examples may be seen in Figures 11, 44, and 54.) Once guidelines that encode vertices, edges, and incircle subdivisions of faces are placed, it is no longer possible for a viewer to visually discern the original intent of the artist. Yackel imagines most artists to be like herself in that when marking a specific set of guidelines, she envisions both the spherical polyhedron and its dual, and conceives of these polyhedra as equal rather than viewing either polyhedron as the main goal.

Nonetheless, not all temari balls explicitly exhibit polyhedral duality. Indeed, the use of incircle guidelines is not required in temari, and if a temari artist avoids these guidelines then the dual may not be apparent on the resulting ball. Because of the beautiful symmetry offered by incircle guidelines, we next consider their accurate placement.

2.2 Making Mathematically Exact Markings

As mentioned in Section 1, accuracy of guideline placement on a temari ball is extremely important as the symmetry of the final product is integral to its beauty. Most mathematicians are gluttons for symmetry and can also appreciate the principle of trying to plot the vertex points in some pure sense, rather than using approximation. Here we will give a mathematically pure construction for point plotting, while being mindful that a physical sphere is not ideal and so our best efforts are rendered approximations. Readers uninterested in exact constructions will unfortunately therefore not be able to appreciate the mathematical aesthetic.

Our methodology is analogous to straightedge-and-compass construction. The primary tools are flexible paper tape and straight pins. Paper tape is used to measure distances on the surface of the purported sphere, and pins are used to mark points. We allow the tape to swivel about a point on the ball by inserting a pin into the tape at a point and then moving the tape freely about this point. In addition, we allow the use of both origami techniques and straightedge-and-compass procedures on the paper tape. For example, we can divide the paper tape exactly into any number of subdivisions using straightedge-and-compass techniques. Alternatively, we can use the origami techinique of Fujimoto approximation (see [7] for an excellent discussion) to approximate this result to the desired degree of accuracy.

The guideline marking process begins with measuring the circumference of our ball using paper tape. We then mark a point (with a pin), use the tape to determine its antipodal point, and viewing those two points as poles, mark any desired number of equally spaced points around the corresponding equator.

Wrapping a guideline is much like drawing a line. In order for a guideline not to slip, one usually requires four points, spread fairly evenly, for a great circle, or three points for half a great circle. Mathematically, two points determine a line, so it is important for us to ascertain that our three or four points lie on the same given great circle before using them to wrap. Detailed instructions for using this methodology to wrap guidelines for three particular temari are given in Section 4. In the next two sections, we describe the general procedures for locating guidelines for the Platonic solids and their truncations.

2.3 Exact Marking for Spherical Platonic Solids

We begin with the octahedron. Having marked two poles and an equator as in Section 2.2, we may then fix four equally spaced points along this equator using the circumference-length paper tape. Together with the poles, these four points can be thought of as the vertices of a spherical octahedron. Adjacent vertices are one quarter-circumference apart as measured along the sphere. The midpoints of the octahedron edges can be easily obtained after dividing the paper tape into eighths.

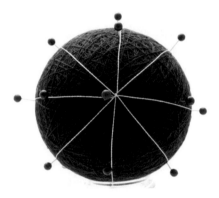

Figure 10. Wrapping guidelines from a pole divides the ball into sectors.

Four guidelines are now wrapped from a pole, dividing the ball into eight orange-like sectors. See Figure 10.

Notice that marking an even number of equally spaced points along the equator ensures that the points fall into antipodal pairs. Hence, any great circle through one of those points passes through the other as well. That is, the sectors are most easily wrapped by starting at a pole, wrapping down to an equator point, down to the opposite pole, up to the antipodal equator point, back to the initial pole, rotating the ball clockwise by 45°, and repeating until all the sectors have been made. This process should then be repeated with the remaining two antipodal pairs of octahedral vertices. At this stage the guidelines triple intersect at the centers of the octahedral faces, forming the faces' incircle subdivisions and marking cube vertices, as shown in Figure 11. This method for marking the octahedron and cube is used universally by temari artists.

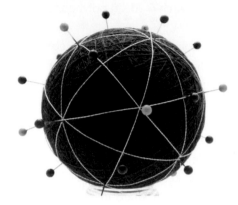

Figure 11. Completed cube/octahedron guidelines, with triple intersection points indicating cube vertices marked with green pins.

The most straightforward mathematical method for marking vertices of the spherical tetrahedron begins with marking the octahedron and cube. (Some temari authors note that the four tetrahedron vertices are equidistant and espouse a guess and check method. However, that cumbersome method requires many steps.) Select four nonadjacent cube vertices to inscribe the regular tetrahedron in the cube, as shown in Figure 12.

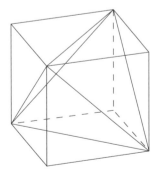

Figure 12. The tetrahedron can be inscribed in a cube.

A much more convoluted but mathematically similar idea can be employed to plot the vertices of a spherical icosahedron based on the fact that a cube can be inscribed in a dodecahedron, as shown in Figure 13. Yackel developed such a method in her quest for a pure solution to the problem of marking icosahedral/dodecahedral guidelines [17]. In response, mathematics educator Mike Naylor suggested the more elegant (and easier!) solution of plotting dodecahedral vertices rather than icosahedral vertices. Yackel then interpreted and developed that idea for the medium of temari as presented here, and we strongly urge all stitchers to adopt the resulting Twenty method. (Detailed instructions are given in Section 4.2.)

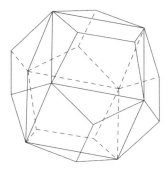

Figure 13. A red cube inscribed in a black dodecahedron.

The Twenty method begins by wrapping a cube/octahedral subdivision, and determining the side length, ℓ, of the spherical cube. In addition, notice that the eight vertices of the cube are also vertices of the dodecahedron. To plot the remaining 12 vertices of the dodecahedron, we must locate points a distance ℓ from our existing dodecahedral vertices. Finding points that are at a distance ℓ from pairs of cube vertices produces equilateral triangles; however, we must choose these new points so that any two such points within adjacent cubical faces are *also* a distance of ℓ apart. This will assure that all 20 plotted points are equidistant.

To carry out this procedure, project Figure 13 to a sphere and note that the desired points lie on nondiagonal bisectors of cubic faces. Measure ℓ from a cubic vertex onto such a bisector, and plot the point obtained. Next measure from another cubic vertex to obtain another point. These two points should each be doubly determined. That is, two different vertices produce each point. (See Figure 45 in Section 4.2.) The great circle extension G of the selected bisector also runs through two dodecahedral vertices on the opposite cubic face. Next note that on the cubic faces adjacent to the originally selected face, the dodecahedral vertices lie on the bisector perpendicular to G. This completely determines the plotting of the remaining dodecahedral vertices. Incircle guidelines can now be wrapped to finish the dodecahedral/icosahedral subdivision.

Finally, we shed mathematical light on the most accurate commonly used method for marking a temari dodecahedron/icosahedron. This approximates spherical icosahedral edge lengths by $\left(\frac{1}{6} + \frac{1}{100}\right)$ times the circumference of the ball. Why this "magic" fraction, as it is seen by many in the temari community? Consider a triangle determined by adjacent vertices of an icosahedron and the center of its circumscribing sphere (shown in Figure 14).

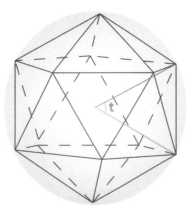

Figure 14. An inscribed icosahedron with central angle labeled t.

We wish to measure the arc of the sphere corresponding to the angle labeled t, i.e., the radian measure of t times the radius. Conceive of the icosahedral vertices as vectors in the standard coordinatization $\{(0, \pm 1, \pm \phi), (\pm 1, \pm \phi, 0), (\pm \phi, 0, \pm 1)\}$ (where $\phi = \frac{1+\sqrt{5}}{2}$ is the golden mean). Each vector has length $\sqrt{(5+\sqrt{5})/2}$, and each pair of vectors has dot product ϕ. (For example, check $(1, \phi, 0)$ and $(-1, \phi, 0)$.) Recall that for any vectors u, v, the formula $u \cdot v = \|u\| \, \|u\| \cos t$ holds. Solving gives $t = \cos^{-1}(\frac{\sqrt{5}}{5}) \approx 1.107148718$ radians. Dividing by 2π returns $.176208$ as the proportion of the circumference of the ball between icosahedral vertices. While $\frac{1}{6} = .166\bar{6}$ is a reasonable approximation, $\frac{1}{6} + \frac{1}{100} = .176\bar{6}$ is a significantly better approximation to $.176208$.

2.4 Marking Archimedean and Catalan Polyhedra

After the Platonic (regular) solids, the next natural class of polyhedra to mark on temari balls are the semiregular (Archimedean) solids and their duals (Catalan solids). Semiregular polyhedra have at least two types of regular polygon faces, and each vertex has the same polygons surrounding it in the same order. (This is consistent with the definition of semiregular tessellations given in Section 2.3 of Chapter 9.) There are 13 Archimedean polyhedra [3, pp. 156–168] and thus 13 Catalan solids. All are pictured in Figures 15 and 16, where corresponding entries are dual.

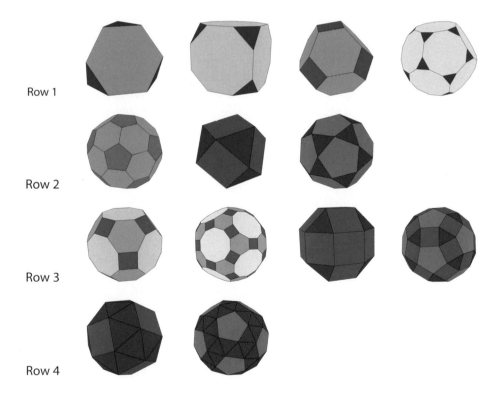

Row 1

Row 2

Row 3

Row 4

Figure 15. The Archimedean solids. Row 1: Truncated tetrahedron, truncated cube, truncated octahedron, truncated dodecahedron. Row 2: Truncated icosahedron, cuboctahedron, icosidodecahedron. Row 3: Great rhombicuboctahedron, great rhombicosidodecahedron, small rhombicuboctahedron, small rhombicosidodecahedron. Row 4: Snub cube, snub dodecahedron.

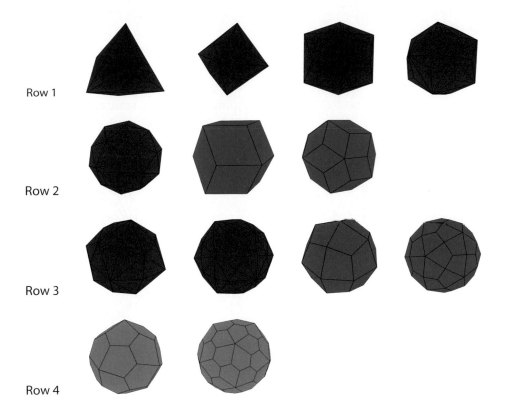

Row 1

Row 2

Row 3

Row 4

Figure 16. The Catalan solids. Row 1: Triakis tetrahedron, small triakis octahedron, tetrakis hexahedron, triakis icosahedron. Row 2: Pentakis dodecahedron, rhombic dodecahedron, rhombic triacontahedron. Row 3: Disdyakis dodecahedron, disdyakis triacontahedron, deltoidal icositetrahedron, deltoidal hexecontahedron. Row 4: Pentagonal icositetrahedron, pentagonal hexecontahedron.

Here we present pure methods for marking incircle guidelines for the seven of the Archimedean polyhedra obtained by truncation, and by Theorem 1 these yield guidelines for the corresponding Catalan solids. *Truncation* is the process of symmetrically slicing off a region near each vertex as shown in Figure 17. For the Platonic solids, truncation depth can be chosen so that the faces of the resulting polyhedra are regular. This gives two choices: a shallow truncation that doubles the number of sides of each original face and a deep truncation that cuts half way down each side, leaving the number of sides of original faces fixed. The shallow depths yield

the first five polyhedra listed in Figure 15. The deep truncation pares duals down to similar solids. That is, the cube and octahedron both truncate to the cuboctahedron, and the icosahedron and dodecahedron both truncate to the icosidodecahedron. The tetrahedron truncates to the octahedron.

To mark guidelines for the deep truncations corresponding to the spherical cuboctahedron and spherical icosidodecahedron (respectively), begin with cube/octahedral (respectively dodecahedral/icosahedral) incircle guidelines and pass additional guidelines through the midpoints of adjacent pairs of edges. This marks the vertices of the spherical cuboc-tahedron (respectively spherical icosidodecahedron). Complete incircle guidelines from this set of vertices.

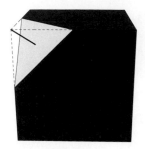

Figure 17. Truncation.

Marking a shallower truncation that produces all regular faces initially appears more challenging. For ex-ample, it would be problematic to determine polyhedral side lengths and then project because radial projection is not an isometry. Instead we will bisect angles, which are preserved under radial projection. The effect of shal-low truncation on a single face of a Platonic solid is to input a regular polygon, excise the corners, and return a regular polygon with twice as many sides. The new polygon can be formed by bisecting the central angles of the initial polygon and using these angle bisectors to

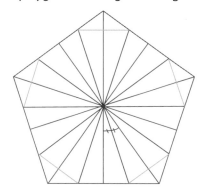

Figure 18. New edges from truncating a face of a Platonic solid are shown in green.

determine the vertices of the new polygon, as shown in Figure 18. Joining the endpoints of adjacent bisectors with line (guideline) segments completes the projected polygonal face. These endpoints form the vertices of the truncated Platonic solid, as depicted in Figure 19. Finally, incircle guidelines can be completed from this new set of vertices.

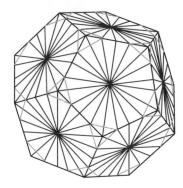

Figure 19. A truncated dodecahedron. New edges are green.

The above theoretical methods for marking these seven Archimedean/Catalan solid pairs do not directly convert to efficient temari guideline laying instructions. A straightforward method for laying temari guidelines for the truncated cube/small triakis octahdron pair is given in Section 4.3. It is an open problem to convert the given theoretical methods into efficient directions for laying guidelines, and to mark vertices for the remain-ing Archimedean polyhedra and their Catalan duals.

2.5 Spherical Symmetry Types of Temari Balls

Our focus thus far has been on examining and produc-ing guidelines for highly symmetric temari balls. Now we turn to a study of algebraic symmetry types of spheres. For each of the 14 discrete spherical symmetry

Figure 20. From left to right, symmetry types *532 (dodecahedral/icosahedral), *432 (cubic/octahedral), and *332 (tetrahedral). Diagrams by Chaim Goodman-Strauss [2].

Figure 21. From left to right, symmetry types *532 (dodecahedral/icosahedral, pattern from [13, p.58]), *432 (cubic/octahedral, pattern from [8, p. 59]), and *332 (tetrahedral, pattern from [11, p. 34]).

types, we give an example of a temari ball of that type. This shows that each of the spherical symmetry types can be instantiated in temari. We adopt the notation used in [2] and refer the reader to this source for the theory behind the classification of spherical symmetry groups.

A majority of the given temari examples are new, and in fact most published temari patterns represent a minority of the 14 classes. We will say more about this in Section 2.6, but for now, we point out that each Platonic, Archimedean, and Catalan solid has one of the first three spherical symmetry types listed here.

The spherical symmetry types derived from projected Platonic solid dual pairs may have or lack reflection symmetry within each polyhedral face. The three types with reflection symmetry are shown in Figure 20; notice that the region marked on each sphere matches the smallest region delineated by incircle guidelines as in Figures 11 and 44. Temari balls exhibiting these sym-

metry types are shown in Figure 21. Figure 22 shows the three spherical symmetry types with Platonic faces that have no reflection symmetry. Temari balls exhibiting these chiral symmetry types are shown in Figure 23. The chirality in each comes from the weaving of the embroidery.

Next we exhibit the seven spherical symmetry types derived from the seven frieze patterns (see [1] or [2] for explanation) by wrapping them around sphere equators. These are shown in the abstract in Figures 24, 26, and 28 and instantiated on temari balls in Figures 25, 27, and 29. The basic guidelines for these consist of a number of great circles dividing the ball into sectors, as in Figure 10, together with an equator. The remaining temari designs in this section were developed by Yackel.

In Figure 25 (left), the white flower appears only on one side of the ball to eliminate pole-reversing symmetry, and though it cannot be seen in a single image, Figure 25 (right) is not top/bottom symmetric.

Figure 22. From left to right, symmetry types 532 (dodecahedral/icosahedral), 432 (cubic/octahedral), and 332 (tetrahedral). Diagrams by Chaim Goodman-Strauss [2].

Figure 23. From left to right, symmetry types 532 (dodecahedral/icosahedral, pattern from [13, p. 56]), 432 (cubic/octahedral, pattern from [9, p. 71]), and 332 (tetrahedral, pattern from [11, p. 34]).

Figure 24. Spherical symmetry types *NN (left) and NN (right). Diagrams by Chaim Goodman-Strauss [2].

Figure 25. Spherical symmetry types *NN as *11 11 (left) and NN as 17 17 (right).

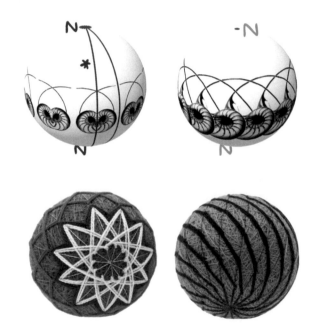

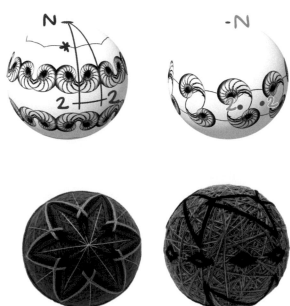

Figure 26. Spherical symmetry types *22N (left) and 22N (right). Diagrams by Chaim Goodman-Strauss [2].

Figure 27. Spherical symmetry types *22N as *226 (left) and 22N as 225 (right).

Figure 28. From left to right, spherical symmetry types N*, 2*N, and N×. Diagrams by Chaim Goodman-Strauss [2].

Figure 29. From left to right, spherical symmetry types N* as 4*, 2*N as 2*7, and N× as 4×.

Figure 30. Spherical symmetry type 3*2. Diagrams by Chaim Goodman-Strauss [2].

Figure 31. Spherical symmetry type 3*2 on a temari ball.

Although only one side of the ball is shown in Figure 27 (left), it does have symmetry over its equatorial plane. Symmetry over this plane is precluded in the ball shown in Figure 27 (right) by the fact that the top and bottom stars are chiral but not mirror images. The diamond shapes along the equator add visual interest, but do not contribute to the symmetry class.

The motif on the hemisphere opposite to that shown in Figure 29 (left) is a mirror image of that pictured, so this ball again has reflection symmetry over its equatorial plane. The motifs on the ball shown in Figure 29 (right) are freehand stitched with counterclockwise spirals on the upper half and counterclockwise on the lower half.

The remaining spherical symmetry type is shown in Figure 30; it includes a projected octahedron but not a projected cube. Each octahedral face has rotational symmetry and each octahedral edge has reflective symmetry.

Figure 31 shows this symmetry on a temari ball. Notice the reflective symmetry exhibited by the reversal of the weaving direction as one moves across octahedral edges marked by metallic guidelines.

2.6   Classifying Temari Ball Patterns

Many temari ball patterns are based on the Platonic solids, as should not surprise the reader of this chapter. Yet all temari patterns based on Platonic, Archimedian, and Catalan solids fall into one of the first six spherical symmetries listed in Section 2.5. Therefore, the 14 spherical symmetry groups do not sufficiently distinguish between temari balls. For example, the truncated cube ball is visually very different from a cubic/octahedral ball (compare Figures 50 and 34), yet both have symmetry type *432. Moreover, there are temari artists who use spherical polyhedral subdivisions that correspond to polyhedra not considered in this chapter, such as spherical buckyballs and other truncations. The resulting balls display stunning feats of embroidery to tremendous effect (see [11]) and are visually distinct from the temari balls depicted here with which they are nonetheless classified via the algebraic system described in Section 2.5. Although group theory is in itself beautiful it does not give a fine enough classification system for design (as noted in [5, p.22]).

How then shall we distinguish between different temari balls? We suggest at a minimum classifying temari ball patterns by the (often dual pairs of) projected polyhedra underlying their stitching together with an accounting of the symmetry or chirality present in facial motifs. By taking into consideration the motif structure, this notion is similar to the algebraic system described in Section 2.5; yet, tracking the underlying polyhedral structures assures that any such classification would result in a refinement of the 14 spherical symmetry group types.

3   Teaching Ideas

The making of temari balls by students allows the teacher to introduce or reinforce a great variety of topics, as explained below. Moreover, even a poorly made temari ball is usually beautiful. Instructions for two very simple balls, one exhibiting cube/octahedral symmetry and one exhibiting dodecahedral/icosahedral symmetry, are given in Section 4. With any hands-on activity, some students will want or need one-on-one assistance and therefore we recommend a maximum student:teacher ratio of 12:1; one way to achieve this is to recruit helpers for the activity. You and the helpers should each have laid the guidelines for the temari ball you are instructing on three or four different styrofoam bases, as it takes practice to compensate for the unique failure of each ball to be a sphere. Instructors are encouraged to have the students collaborate and to monitor the collaborations for correctness.

On the purely practical side, it is important to plan how to allocate class time to making temari balls. The following times are given for a mathematics for liberal arts class at a comprehensive university. First students must choose base thread color and three or four coordinating perle cotton colors. Yackel buys spools of serger thread and large packages of "craft thread" (which is really perle cotton) specifically for her classes. Giving the students five minutes to pick out their materials from well-distributed centers around the room as the last task before leaving for the day allows indecisive folk to pon-

der for up to 12 minutes before the next class needs the room.

Putting a base wrapping on a temari ball takes about 20 minutes, and students should be given instructions and then attempt this at home to save class time. (They will be distracted if attempting to wrap during other mathematical activities.) Even sloppy wrapping jobs can be given a final layer of tight wrapping at the start of the next class.

Plotting points on a first cube/octahedral ball requires 30 to 40 minutes of class time. Wrapping guidelines then takes a class of 25 about 20 minutes. Placing pins for the first motif is best done as a class so that all students have a common understanding. Next the students must master the herringbone stitch. At this point the instructor must simply check that students are stitching right-to-left while rotating the ball clockwise (or stitching left-to-right while rotating counterclockwise). Simultaneously, problems of not stitching "under" the pins, under the appropriate guidelines, and under all threads on the succeeding rounds can be nipped in the bud. Starting from pin placement, the process of completing at least one cube/octahedron motif should take half an hour. Having helpers to check work will reduce the time. While not mathematical, this in-class time use will pay off as no other stitching need take place in class on this ball or the next.

Plotting points for a second dodecahedral/icosahedral ball using the $(\frac{1}{6} + \frac{1}{100})$ method without explanation requires about half an hour. There is a significant variation in student completion times for wrapping dodecahedral/icosahedral guidelines. (A few students essentially require the instructor to have her hands on their hands while wrapping this second ball, which is very difficult for all parties.) Because this can easily take a whole 50 minute class period, we recommend this be made a take-home exercise (with directions and hints!) followed by an in-class discussion and intervention. During the in-class portion, students can help each other, and students who need excessive help can be directed to office hours.

We now examine the mathematical questions that arise during temari ball construction. While plotting points, students at the middle-school level or higher should consider:

★ How can one plot a pair of antipodal points?

★ How can one plot an equator for those antipodal points?

★ How can one make a set of n equally spaced sector lines for a pair of antipodal points for any given n?

Common student conceptual errors include not realizing that antipodal points must be half a circumference apart, measuring the ball circumference loosely, and marking a circle equidistant from one pole point that is not an equator. Measuring from the endpoints of a segment on the equator to a pole creates an angle at the pole. Equal equator segments correspond to equal pole angles; it is worth eliciting this idea from students, but they are also likely to generalize it inappropriately.

Wrapping guidelines is fun but usually difficult at first. Here are some questions for students to consider; some questions are easier to answer for those with more guideline-wrapping experience.

★ What do the regions enclosed by the guidelines look like as you wrap? (The answer to this question changes as you wrap.)

★ How many guidelines go through each vertex when you are done wrapping? How many strings appear to emanate out of each vertex? And, how should the two answers be related?

★ Do the vertices cluster into groups? How many groups? How many are in each group? Do these numbers remind you of anything? (Hint: think of the Platonic solids.)

★ What do the regions enclosed by the guidelines look like when you are done wrapping? (What do you think the answer is supposed to be? This might suggest that you adjust the guidelines on your ball to make it so.)

★ If you think about the Platonic solids, is there a subset of the vertices and a way of combining regions that will make your guidelines look like you put one of the Platonic solids on the sphere? Is there another subset of the vertices that will work similarly? If so, how are the resulting Platonic solids related?

★ In making the guidelines, how many great circles did you make? (Don't count the partial great circles made in moving from one pole pair to another to complete the next set of sector lines.) How many did you need to add when you wrapped each new pole pair?

Interesting mathematical points can be introduced during the stitching phase.

★ When a motif is stitched on a temari ball, the ball is rotated. For example, between stitches of a four-pointed polystar, the crafter rotates the ball by 45° about the polystar center. This motion can be thought of as a group generator for $\mathbb{Z}/(8)$. Why does stitching a petal (stitching far away from the vertex) on every other guideline lead to exactly four petals rather than a big mess of petals? For example, what would happen if nine sector lines emanated from the center vertex and one stitched a petal on every other guideline?

★ In a similar vein, when stitching the triangles around the cube vertices of the 6/8 ball, by what angle do you rotate? What cyclic group does this generate?

★ If you have made the 12/20 ball, explain the stitching of the three- and five-pointed flowers in terms of modular arithemetic, as in the previous question.

3.1 Project Ideas

Readers of Chapter 6 may recognize the stacked flowers shown in Figure 34 as representing cosets of a subgroup of D_4. Design a cube/octahedron temari ball with different faces representing different cosets. Because there are seven proper subgroups and only six cubic faces, it will be interesting to choose which subgroup to omit. Temari techniques not discussed here will be needed to properly express the nuances of the cosets, so consult [10] for assistance.

Some questions arise if one wishes to conserve perle cotton. When is it best to cut the yarn and rebury the end (a pesky process) rather than simply stitching to the next closest point? In what order should motifs be stitched so that one does not end up with isolated (unstitched) motifs at the end? How does the situation change if one allows the temari artist to partially stitch a motif, move to stitch another motif, and return later to complete the first motif? A practical mathematical solution would have to be very easy to follow so that stitchers could make use of it without ruining the enjoyment of the craft. Approaching these open-ended questions allows the instructor to discuss principles of logical thinking, distance, and graph theory with the students.

4 Crafting Three Temari Balls

Each temari ball in this section demonstrates a particular spherical subdivision: the first is of the cube/octahedron, the second is of the dodecahedron/icosahedron, and the third is of the truncated octahedron. One of the dodecahedron/icosahedron marking methods given is original, as are the truncation instructions and marking for the truncated octahedron. Temari artists are highly encouraged to try out the elegant Twenty method as it gives exact, rather than approximate, point placement for the dodecahedral/icosahedral subdivision and should overcome the perennial problem of irregularly sized and shaped pentagons! The first two patterns are deliberately plain in order to emphasize underlying structure. Experienced temari artists are encouraged to embellish, given the basic guideline subdivisions suggested, to obtain a variety of new looks. Beginners desiring additional assistance are directed to the Internet resource [12] or any of the English temari books [4], [10], [14], or [15]. (The temari books published by Macaw are all written in Japanese.) The directions that follow are presented in a linear fashion and later directions often reference instructions for previous balls. Therefore, beginning temari stitchers are encouraged to attempt the projects in order.

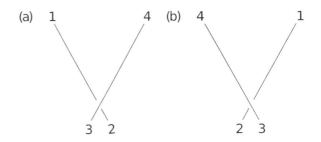

Figure 32. How to make a herringbone stitch. Bringing the yarn from the previous stitch at position 1, make the herringbone stitch by inserting the needle at 2 and bringing it back out at 3. Continue on to the next stitch at 4. Right handers use (a) and continue working toward the right; left handers use (b) and continue working toward the left.

The *herringbone stitch* is the only kind of embroidery stitch used in this chapter; see Figure 32 for instructions. A herringbone stitch produces a decorative overlap at the stitch points, as shown in Figure 33.

The length of a herringbone stitch is often quite small. However, these stitches are usually to be taken around a specific location, such as a guideline. In this case, the stitch should be taken underneath and around the guideline. Also, every stitch should be taken under some wrapping thread. Sometimes the stitches slip, which is why one should avoid parallel wrapping threads and instead wrap in all different directions.

Figure 33. A stitch taken close to the center of a star shaped motif (left) and one taken far from the center (right).

Usually temari are designed to have rows of perle cotton strands aligned all the way down to the stitch tips. This is achieved by taking herringbone stitches in one of two different ways, as shown in Figure 33. When working away from the center of a motif at the bottom of a point, as in Figure 33 (right), take a small herringbone stitch. Leave as much as $\frac{1}{16}''$ or even $\frac{1}{8}''$ gap between stitches, as the gap will disappear because of the angle of the stitch and because of slippage. When working at the inside of a point, in Figure 33 (left), make your herringbone stitch directly below the previous stitch; slipping is not a problem in this case. This can mean taking a comparatively long stitch. In addition, if the stitch is long, passing the needle under wrapping threads may not be necessary. Finally, when you finish a motif, stitch under the wrapping thread to move your needle location to the next motif. In either case, be sure to stitch farther away from the center of the motif than all previous layers of perle cotton, as shown in Figure 33.

Note: While embroidery floss is easily available and thus a tempting substitution for perle cotton, stitchers will have difficulty achieving the crisp look of aligned stitches when using it, so we advise against this substitution.

4.1 The Cube/Octahedral or 6/8 Ball

The temari ball in Figure 34, for which we give directions in this section, is nicknamed the 6/8 ball because the guidelines mark off regions for the dual Platonic solid pair the cube (six faces) and the octahedron (eight faces). For an explanation, see Section 2.1.

Figure 34. A completed 6/8 ball.

Figure 35. Three 6/8 temari balls, in perle cotton (left, right) and embroidery floss (center).

Materials

* Three dozen straight pins with large plastic heads in a variety of colors (at least 6 in red, 12 in black, 8 in green, and 4 in yellow).

* A piece of paper (quilling strips are convenient).

* Scissors for paper and for thread.

* A darning or embroidery needle (at least $1\frac{1}{2}''$ long, with an eye big enough for perle cotton).

* A 3'' diameter styrofoam ball.

* 300+ yards of wrapping thread. This should be the weight and thickness of sewing thread and can be made of any fiber.

* 90'' (2.5 yards) of guideline thread. This must be a contrasting color to the wrapping thread, and could be a single strand of metallic embroidery thread such as Mouliné DMC or even a full strand of thicker metallic banding.

* Three coordinating colors of craft thread or perle cotton (such as DMC), less than one hank each.

Notes

All of the threads (wrapping, guidelines, and craft) should look nice together to your eye. Possibilities are shown in Figure 35. Think about what part of the design you would like to have stand out and choose colors accordingly. Changing which colors are bold and which are subtle can completely change the appearance of a design.

Instructions

1. *Wrap the styrofoam ball with the wrapping thread.* Simply begin wrapping the thread around the ball tightly so that the thread is smooth against the surface of the ball and does not sag or slip. Change directions frequently so that the threads *do not* line up next to each other in rows but rather go every which way. Wrap the ball until none of the styrofoam shows. The thread should give a little bit when the ball is squeezed. (Do not squeeze so hard as to smash the styrofoam core.) If you have wrapped the thread very tightly, you may not feel the give; in this case, stop when there is approximately a one millimeter layer of thread covering the ball. When you are done, cut the thread. Thread a needle with the end of the thread, and make several stitches through the thread wrapping on the ball to secure the thread end. Cut off the last little bit of thread that you will never be able to get to stay underneath the thread wrapping. You are now done wrapping the ball.

Figure 36. Measuring and marking the circumference and south pole.

2. *Place the north and south pole pins.* Cut a thin (less than 1 cm wide) strip of paper that is about 11″ long or use quilling paper. Pin one end of the paper into the ball with a red pin. That pin marks the north pole. Using the paper, measure the circumference of the ball as follows.

Place a pin through one end of the paper tape and wrap the tape around the sphere, making a fold at the pin. If the paper constitutes a great circle at this point, then this is the longest the paper will be and still lie flat around the ball. If not, when swiveled around the ball, there will be some other position at which the tape will need to be longer in order to reach the pin while lying flat around the ball. In other words, in order to check that the paper is the right length from pin to fold, turn the ball around the pin, holding the paper tape fixed and reflattening the tape as the ball is turned. If the tape continues to just reach the pin exactly, then the initial circumference estimate was correct. However, if the length of tape either overshoots the pin or undershoots it at some points as the ball is turned, then make sure to straighten the tape and refold it to lengthen the distance to the fold so that the new distance just reaches the pin. Because the goal is a great circle, the tape will always

be lengthened, never shortened. *Do not tear the tape at the circumference!* See Figure 36.

Next, you need to plot the south pole. Fold the circumference in half. Place a red pin at the south pole, but don't put it through the tape! (I recommend using the same color pin as you used at the north pole.) Again, measure the distance from the north to south pole by rotating your tape in different directions from the north pole. As you do this, you will undoubtedly want to move your south pole pin. Go ahead and do this, continually rotating and remeasuring until your south pole seems right. The need to reposition the pin is caused by the imperfections of the ball and the fact that when you placed the south pole pin, there were many options for where to put it along the bottom of the folded paper tape because of the tape's nonzero width. Now that you have put in your north and south poles, do not move them. *In general, once you have carefully measured and placed a pin, you should never move it.*

3. *Place the equator pins.* In this step, we first find the equator by folding the circumference tape in half again, so that it is now divided into fourths. Measuring one fourth of the circumference distance down from the north pole, place at least eight pins, ap-

Figure 37. Measuring one quarter circumference from the north pole (left) allows pins to be placed on the equator (center). The pins can then be repositioned to divide the equator exactly into eighths (right).

proximately equally spaced and alternating red pins with black pins. Next, make a cut through the paper tape to the pin at the north pole in order to remove the tape from the pin without shortening the paper. Fold the circumference tape into eighths, marking the eight points in pen for readability. Align the tape underneath the equator pins to form a circumference, as shown with black pins wearing green pajamas in Figure 37. Reposition the equator pins, this time dividing the equator into eight equal sections.

Figure 38. Marking points on the "other" equators.

4. *Place the other pins.* Reconceptualize the ball as having a different pair of north and south poles, where the pair of poles are chosen from two opposite (antipodal) red equator pins. Using the initial pole pair

and the remaining red equator pins, position the paper tape around the new equator, as shown in Figure 38. Be sure to align the beginning and ending of the circumference markings with one of the pins, which should ensure that every other eighth (i.e., every fourth) of the circumference aligns with one of the pins. Pin down the tape. Now place black pins at the other four eighth markings around the new equator. Repeat the process with the remaining pair of red pins on the initial equator.

5. *Wrap the guidelines.* To begin any thread on a temari, bury the thread end within the wrapping thread. For perle cotton, this consists of entering the wrapping layer with the needle an inch or two from the starting point, and then exiting at the starting point. For thinner thread (such as a single strand of metallic embroidery floss), take long stitches in several directions, all underneath the wrapping layer, and have the needle finally emerge at a pole point. Pull on the floss to make sure it is firmly embedded. Note that only the end of the floss should be secured in the ball; approximately 88″ should emerge precisely at the pole pin ready to be wrapped around the ball.

The next goal is to tightly wrap four great circles passing through both poles and one pair of opposing equator points, for each pair of poles. From the perspective of a pole, doing so divides the ball into eight equal sectors, looking something like an orange. See Figure 39.

Figure 39. The first set of guidelines divides the ball into sectors.

Starting at a red pin (north pole), wrap a full great circle guided by the pins on the equator for that pole and the antipodal pin (south pole), pulling taut as you wrap. Rotate the ball 45° clockwise when the north pole is reached and wrap another great circle. Repeat to obtain four great circles at the pole that divide the ball into eight sectors. Be sure to take a small securing stitch with guideline thread as you wrap past the south pole on the final great circle, and take another securing stitch at the north pole when you complete the eight sectors. A securing stitch is a very small stitch going underneath all the guideline threads and through some of the wrapping thread; it will pull the guideline threads into place and keep them from moving out of place when the pin is removed. While the goal of the stitch is to ensure that the guidelines pass through the point marked by the pin at the pole, it is also supposed to be invisible, so long or many stitches would be antithetical to the purpose. Such securing stitches should be made at *every* pole point, north and south, on the ball.

Bury the guideline and cut, or use the needle to pull under the wrapping thread layer to one pin of the next pole pair. Repeat at each north/south pole pair.

When this division is completed for the three pole pairs, the sector lines cross. Indeed, at eight points on the ball, three sector lines cross. If they don't cross at one point, roll them slightly until this happens, and then place a green pin at each of the eight three-way crossing points.

6. *Stitch the first four-pointed star at the red pins.* The following two-colored four-pointed star will be embroidered at each red pin. Measure $\frac{3}{4}''$ down the guideline from the red pin toward each black pin. Place a new pin (of any color, we'll call it yellow) precisely at this point on the guideline. You should have placed four new pins. Remove the red pin and the four black pins for convenience. Now make a four-pointed star around the center point (where the red pin was) by making herringbone stitches alternately at the yellow pins and the intermediate guidelines close to the center point about the next guideline as shown in Figure 40, turning the ball 45° about the center point

Figure 40. Stitching a four-petaled polystar: right-handed stitchers will follow the stitch order 1–9 (left) to make three rounds of color 1 (center) and then two rounds of color 2 (right).

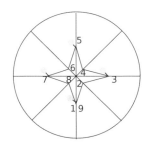

Figure 41. Pin placement for the second four-petaled polystar (left) and the completed polystar (right).

between taking stitches. Such a stitching pattern that alternates between spoke pins and guidelines near a center point is called a *polystar* pattern. Make three rounds with color 1 and two rounds with color 2. Remove the yellow pins. Repeating this for each red pin will result in six two-colored four-pointed stars.

7. *Stitch the second set of four-pointed stars.* Using the same center points as in step 6, again make four-pointed stars. However, this time the star points will be on the other four guidelines as shown in Figure 41. Mark the location for stitching the star points by measuring $\frac{1}{4}''$ toward the center of the star from the green

pins and placing yellow pins at these locations. Stitch two rounds in color 3 about each center to make six four-pointed stars. For ease of stitching you will need to remove the green pins, but please replace them afterwards for reference in future steps.

8. *Stitch the triangles around the green pins.* Six guidelines emanate from each green pin. Place pins $\frac{3}{8}''$ from each green pin along the three guidelines used by the second set of four-pointed stars. With the herringbone stitch, embroider triangles using two rounds of color 2 followed by three rounds of color 1. See Figure 42.

Figure 42. Pins and stitching for the triangles (left), stitching to a new motif (center), and a completed triangle (right).

Figure 43. Twelve stars and twenty triangles make up the 12/20 ball.

4.2 The Dodecahedral/Icosahedral or 12/20 Ball

The temari ball shown in Figure 43, for which we give directions in this section, has the nickname the 12/20 ball because the guidelines mark off regions for the dual Platonic solid pair the dodecahedron (12 faces) and the icosahedron (20 faces). For an explanation, see Section 2.1.

Materials

* Three dozen straight pins with large plastic heads in a variety of colors (at least 6 in red, 12 in black, 8 in green, and 4 in yellow).

* A piece of paper (quilling strips are convenient).

* Scissors for paper and for thread.

* A darning or embroidery needle (at least $1\frac{1}{2}''$ long, with an eye big enough for perle cotton).

* A ruler with centimeters.

* A 3″ diameter styrofoam ball.

* 300+ yards of wrapping thread. This should be the weight and thickness of sewing thread and can be made of any fiber.

* 196″ (5.5 yards) of guideline thread. This must be a contrasting color to the wrapping thread, and could be a single strand of metallic embroidery thread such as Mouliné DMC or even a full strand of thicker metallic banding.

* Three coordinating colors of craft thread or perle cotton, such as DMC, less than one hank each.

Choose a pleasing combination of threads (wrapping, guidelines, and perle cotton); possibilities are shown in Figure 43. Two methods of placing the guidelines are given here, the standard method and the Twenty method. The standard method is given as a courtesy for teachers. We recommend the Twenty method to temari artists.

Instructions for Standard Method

1. *Wrap the styrofoam ball with the wrapping thread.* See step 1 of Section 4.1.

2. *Place the north and south pole pins.* See step 2 of Section 4.1.

3. *Place the equator pins.* Details are given in step 3 of Section 4.1. In step 2, the paper tape was marked with the circumference. Divide the circumference into tenths, and mark off these tenths on the tape. Mathematical purists may want to use a ruler and compass

construction; folding enthusiasts may want to do a Fujimoto approximation, which is explained particularly well in [7]. Temari artists typically measure the circumference in centimeters, divide by 10, and use this to mark the paper tape. Place ten equally spaced pins around the equator.

4. *Divide the ball into ten sectors.* As with the 6/8 ball, secure the guideline thread and have it emerge at the north pole. Wrap five great circles through the north and south poles, guided by the ten equator pins. Be sure to make securing stitches as the fifth great circle through each pole is completed, as explained in step 5 of Section 4.1.

5. *Place the remaining pins.* Calculate $(\frac{1}{6} + \frac{1}{100})$ times the circumference (call this distance d) and mark it off on the paper tape. Move five equator pins d away from the north pole, one on each of five alternating guidelines. Now move the other five equator pins, d away from the south pole. These ten pins are the remaining centers of the dodecahedron faces.

6. *Finish wrapping the guidelines.* The next goal is to wrap so that five great circles pass through each of the ten new pins. Secure the guideline thread and have it emerge at one of the ten pins. Focus on this pin and also be aware of the pin directly opposite it on the ball (its antipodal pin). Complete the five great circles passing through these two pins. The guides are the other pins, which are not of uniform distance from the current pole pins in terms of the great circle. Wrap the guidelines carefully, always pivoting in the same direction at the initial pole point, and taking securing stitches as the five great circles are completed at a pole. If you get confused, don't hesitate to release the guideline thread and unwind. Rewrapping the guidelines is not the end of the world. You will gain a better understanding of the process and achieve a better ball in the end. When wrapping is complete,

guidelines should meet in threes, as in Figure 44. As in step 5 of Section 4.1, some guidelines may need to be rolled slightly so that these meeting points occur.

Proceed to step 7 on page 179.

Figure 44. The result of guideline wrapping before removing pins.

Below we present the Twenty method discussed in Section 2.3 above.

Instructions for the Twenty Method

1. *Wrap the styrofoam ball with the wrapping thread.* See step 1 of Section 4.1.

2. *Place pins and then wrap the guidelines for a 6/8 subdivision, but do not take small stitches to secure the guidelines.* Details are given in step 2 of Section 4.1.

3. *Mark off the length of a projected cube edge on a measuring tape.* See Figure 45 (left) for a geometric depiction of the desired length ℓ. Be sure to measure several cube edges and roll guideline intersection points if necessary to make sure you have a length that works for all edges of the cube.

4. In addition to the eight guideline intersections corresponding to the vertices of the projected cube

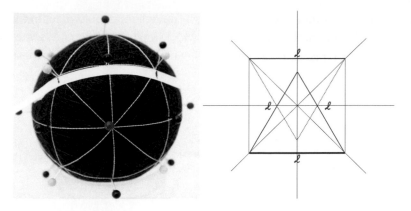

Figure 45. Determining the projected cube edge length (left). Finding remaining vertices (right).

(which are also vertices of the projected dodecahedron), *mark the remaining twelve vertices of the dodecahedron with twelve pins in three different colors.* Begin by selecting a face of the cube. Note that this face has a guideline running through it from top to bottom. Measure ℓ from a face vertex onto this vertical guideline and place a pin in the first color. See Figure 45 (right). Check the pin placement by measuring ℓ from the cube vertex just across the face to that pin.

The two vertices of the cube face together with the new pin should form an equilateral triangle. (Measuring the height of this triangle can be used to give a shortcut for the rest of the pin placements.) Repeat the process on the other two vertices of the same cube face, using the same pin color. In the end, you will have two pins lying on the same guideline. Using the same pin color, repeat the whole process to obtain two more pins on the same guideline but on the cube face opposite the face you just pinned.

At this time two opposing pairs (four total) of unpinned cube faces remain. For each pair, use a new pin color and place pins as before, using a fresh guideline each time. Though this may seem tricky at first, it is actually simple. After the first set of pins has been placed, one guideline on each face will be off-limits, so if one choice of adjacent vertices yields a pin on an already used guideline, simply swap one vertex with that diagonally opposite.

Figure 46. An icosahedral/dodecahedral temari pinned and ready to wrap guidelines with the Twenty method (left) and the same ball wrapped with 12/20 guidelines (right).

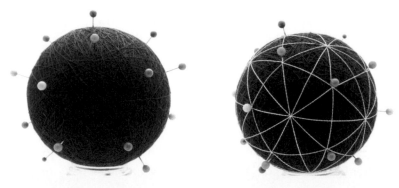

5. *Remove extraneous pins and guidelines.* At this point the ball has all of the 6/8 ball pins. Remove all pins except the eight cube vertex pins and the twelve new pins added in step 4. Next, without moving any of these twenty pins, remove all of the guidelines. (This is why you were not supposed to take any securing stitches in step 1.) From some perspectives the ball will look as in Figure 46, though from others it looks like a bunch of arbitrarily placed pins.

6. *Wrap the guidelines for the 12/20 ball.* Each of the twenty pins represents a vertex of the icosahedron, so three guidelines will pass through it, as in Figure 46. Each of the great circles corresponding to these guidelines is determined by four pins. In order to wrap the guideline set passing through a single pin p, first identify p's antipodal pin $\hat{p}$. Next, find p's three equidistant closest pins, q, r, and s. Starting at p, wrap a great circle through q, $\hat{p}$, $\hat{q}$, and back to p. Now turn and do the same through r and s, noting that small securing stitches may be needed and care must be taken as the angles can be severe. Then move to a nearby unfinished pin and repeat, making only the necessary great circles. Note that some pins may remain untouched until quite late in the process.

7. *Stitch the stars.* A star will be stitched around each of the twelve dodecahedral centers marked with a pin

and at which five great circles meet. Place five pins $\frac{3}{8}''$ from the center on the guidelines that connect dodecahedral centers. Using the herringbone stitch, stitch a star by emerging first at one pin and rotate the ball clockwise, moving past the next pin. Take a stitch at the following pin (the one two away from the first). Continue in this manner until arriving back where you began; this completes a single round of the star. This sequence is shown in the stitch diagram given in Figure 47. Complete a total of three rounds in color 1 so the result is as in Figure 48 (left).

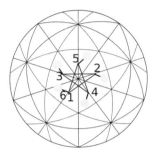

Figure 47. Stitch order for the stars.

8. *Outline the stars.* Using color 2, stitch five-pointed stars around the outside of the stars made in step 8 in a manner similar to step 6 from Section 4.1, except that this time the stitches should be taken close to the star made in step 7, as shown in Figure 48 (right).

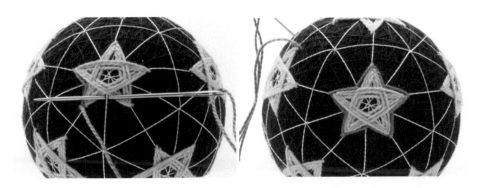

Figure 48. Outlining the stars.

Figure 49. Pinning and stitching a triangle.

9. *Stitch the triangles.* The triangles will be stitched about the twenty points where three great circle guidelines meet. Each meeting point looks like the middle of a larger triangle where the guidelines act as perpendicular bisectors of the angles of the triangle. Place pins $\frac{1}{4}''$ from the base of the triangle on each bisector. Now stitch three rounds in color 3, forming triangles as in Figure 49.

4.3 The Truncated Cube Ball

The temari ball for which we give directions here is a truncated cube, as explained in Section 2.4 above. The stitching on the ball consists only of polystars; the experienced stitcher is challenged to create different patterns based on the given guideline placement.

Materials

* Six dozen straight pins with large plastic heads in a variety of colors (at least 6 in red, 12 in green, 24 in white, 8 in pink, and 24 in yellow).

* A piece of paper (quilling strips are convenient).

* Scissors for paper and for thread.

* A darning or embroidery needle (at least $1\frac{1}{2}''$ long, with an eye big enough for perle cotton).

* A 3″ diameter styrofoam ball.

* 300+ yards of wrapping thread. This should be the weight and thickness of sewing thread and can be made of any fiber.

Figure 50. Four truncated cube balls.

* 450″ (12.5 yards) of guideline thread. This must be a contrasting color to the wrapping thread, and could be a single strand of metallic embroidery thread such as Mouliné DMC or even a full strand of thicker metallic banding.

* Three coordinating colors of craft thread or perle cotton; the ball at the left in Figure 50 uses No. 5 perle cotton, 160″ (4.5 yards) of yellow DMC #742, 250″ (7 yards) of green DMC #367, and 600″ (17 yards) of red DMC #815.

Instructions

1. *Wrap the styrofoam ball with the wrapping thread.* See step 1 of Section 4.1.

2. *Place the north and south pole pins.* Use red for these two pins and see step 2 of Section 4.1.

3. *Place the equator pins.* Details are given in step 3 of Section 4.1 for placing eight equally spaced pins around the equator. Here, place 16 equally spaced pins around the equator in the order red, white, green, white (four times) as in Figure 51.

Figure 51. Place 16 equator points in the correct color sequence.

4. *Place the other pins.* Use the methodology from step 4 of Section 4.1 to place the other pins, but divide each perpendicular equator into 16 subdivisions and use the same sequence of pin colors as in step 3. See Figure 52.

5. *Wrap the guidelines.* Begin wrapping guidelines at a red pin. Wrap all eight great circles and thereby divide

Figure 52. Mark points on the remaining perpendicular equators, dividing each into 16 equal parts.

Figure 53. The ball has been divided into 16 sectors by the guidelines.

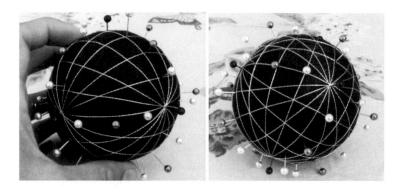

Figure 54. The ball after wrapping the second set of eight great circles (left) and completely guidelined but possibly insanity-producing (right).

the ball into sixteen sectors as shown in Figure 53. Be sure to take a small securing stitch as you wrap past the south pole on the final great circle, and take another securing stitch at the north pole when you complete the sixteen sectors. Next, stitch underneath the wrapping thread to a new north pole. Again wrap eight great circles to obtain sixteen sectors and make securing stitches. Finally, complete the guidelines by wrapping the eight great circles through the third north/south pole pair. See Figure 54. The guidelines should exhibit many three-way intersection points, as shown in Figure 54. Fix errant intersections by rolling

the guidelines until triple intersections occur as they should.

6. *Take stock of the polyhedral faces projected onto the ball.* These will be your stitching palettes, so mark them with pins as indicated in Figure 55. Cube vertices will be marked in pink. The ball also has eight projected triangular faces produced by cutting off the corners of the projected cube, and these will be marked in yellow. However, placing the yellow pins now will make for stitching difficulties, so we suggest adding those pins later. In any case, note that the yel-

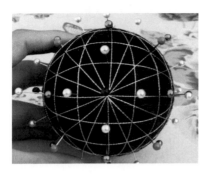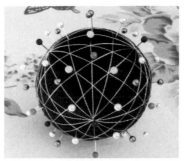

Figure 55. The ball has eight cube vertices marked in pink (left) and eight triangular faces obtained by truncating the corners of the cube, marked in yellow (right).

low pins lie on guidelines connecting pink pins and green pins and surround a pink pin, because a plane that cuts off the vertex of a cube will shorten the cube edges. Take care to mark the correct guidelines to avoid producing upside-down (nonfacial) triangles.

Once all of the yellow pins are in place, they also demarcate six octagonal regions for stitching in place of the former sides of the projected cube.

7. *Stitch the leaf polystars.* Move to an octagonal region. Measure the distance d from the red pin at the center of the octagonal region to the nearby white pins. (Surprisingly, on the sample ball the white pins were exactly $\frac{5}{8}''$ from the red pin.) Number the guidelines (spokes) emanating from the red pin 1 through 16, starting with a white-pin spoke. Place different-colored pins d from the center on spokes 3, 7, 11, and 15; yellow is used in Figure 56. (You may wish to reuse some equator pins at this point.) White pins should already be on spokes 1, 5, 9, and 13. Next, place pins $\frac{15}{16}''$ from the red pin on the remaining eight spokes (2, 4, 6, 8, 10, 12, 14, and 16). Now work three rounds of color 1 in a polystar pattern on this set of pins. Take all polystar stitches far from the center of the octagon, outside both the inner

and outer set of pins, as pictured in Figure 56. Work this pattern in each of the six octagonal regions of the ball.

Figure 56. The leaf polystar.

8. *Stitch the flower polystars.* A flower polystar will be stitched on top of each leaf polystar. On spokes 1, 3, 5, 7, 9, 11, 13, and 15, place pins $\frac{15}{16}''$ from the center red pin. (These are the opposite spokes from those marked in step 7.) Work a four-round polystar pattern in color 2, with alternate stitches taken on even-numbered spokes close to the center. See Figure 57. Work this pattern in each of the six octagonal regions around the ball.

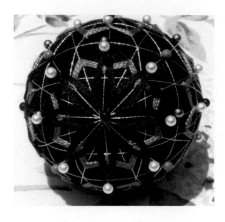

Figure 57. The flower polystar.

9. *Stitch the triangular motifs.* Place the yellow pins, as described in step 6. Use color 3 to stitch a three-round polystar around these three pins, stitching to the center on the alternate spokes. See Figure 58. There are eight such regions on the ball.

Figure 58. The triangle polystar.

Bibliography

[1] belcastro, sarah-marie and Hull, Thomas. "Classifying Frieze Patterns Without Using Groups." *The College Mathematics Journal* 33:2 (2002), 93–98.

[2] Conway, John H., Burgiel, Heidi and Goodman-Strauss, Chaim. *The Symmetries of Things.* A K Peters, Wellesley, MA, 2008.

[3] Cromwell, Peter R. *Polyhedra.* Cambridge University Press, New York, 1997.

[4] Diamond, Anna. *The Temari Book.* Lark Books, New York, 2000.

[5] Grünbaum, Branko. "Periodic Ornamentation of the Fabric Plane: Lessons from Peruvian Fabrics." In *Symmetry Comes of Age: The Role of Pattern in Culture,* edited by D. K. Washburn and D. W. Crowe, pp. 18–64. University of Washington Press, Seattle, 2004.

[6] Grünbaum, Branko and Shephard, G.C. "Duality of Polyhedra." In *Shaping Space: A Polyhedral Approach*, edited by Marjorie Senechal and George Fleck, pp. 205–211. Birkhäuser, Boston, 1988.

[7] Hull, Thomas. *Project Origami.* A K Peters, Wellesley, MA, 2006.

[8] Kanke, Akiko, Shikino, Miyoko, Tomita, Tatsu, and Toyoda, Sakiko. *Dream Temari, Classic to Modern.* Macaw, Tokyo, 1999.

[9] Sato, Ayako, Takahara, Yoko, Tahara, Tomeko, and Momoto, Kayoko. *Creative Temari.* Macaw, Tokyo, 1997.

[10] Seuss, Barbara B. *Japanese Temari.* Breckling Press, Elmhurst, IL, 2007.

[11] Takahara, Yoko. *Flower Temari from the Kaga Region of Kyoto.* Macaw, Tokyo, 1992.

[12] Thompson, G. *TemariKai.com* Available at http://www.temarikai.com/, April 19, 2010.

[13] Tiyoko, Ozaki. *New Temari 5.* Macaw, Tokyo, 1983.

[14] Vandervoort, Diana. *Temari: How to Make Japanese Thread Balls.* Japan Publications Trading Co., Tokyo, 1992.

[15] Wood, Mary. *The Craft of Temari.* Tunbridge Wells, Kent, UK, 1991.

[16] Yackel, C. A. "Embroidering Polyhedra on Temari Balls." In *Math+Art=X Conference Proceedings*, pp. 183–187. University of Colorado, Boulder, CO, 2005.

[17] Yackel, C. A. "Marking a Physical Sphere with a Projected Platonic Solid." *Proceedings of the 2009 Bridges Banff Conference*, edited by George W. Hart and Reza Sarhangi, pp. 123–130. Tarquin Publications, London, 2009.

CHAPTER 9

quilting semiregular tessellations

IRENA SWANSON

1   Overview

There are whole-cloth quilts and patchwork quilts; there are monochromatic quilts and multicolored quilts; there are functional bed quilts and wall quilts; there are treasured never-used quilts and well-used quilts; there are quilts with completely random and nonrepeating patterns and there are quilts with repetitions; and of course there are many more ways of looking at quilts.

There are uncountably many patterns, but life is short, so we cannot make them all into quilts. One has to pick and choose a finite number (first one, then another ...). Below is a systematic reduction of possible patterns to a doable finite number; in the process of this reduction we will learn some mathematical concepts and even prove that the reduction is indeed systematic. If your resulting quilts are not seamed absolutely completely perfectly, at least you can have the satisfaction that you were systematic in choosing the quilt patterns?!

We begin by restricting ourselves to the (smaller) infinite number of quilts that have regularly repeating patterns, and that are composed entirely of regular polygons: a *regular polygon* is a many-sided figure in which all interior angles are the same and all sides have the same length. Every regular polygon is convex, meaning that a needle whose ends are set inside the polygon also has its entire length inside the polygon (see Figure 1).

Figure 1. A convex regular hexagon (left) and a nonconvex hexagon (right).

Among the regularly repeating quilt patterns whose parts are composed of regular polygons we will only examine those in which any two adjacent regular polygons meet exactly edge-to-edge. But even with all of these restrictions, there are still uncountably many different quilts that can be made, even if one does not account for the infinite variety of fabrics, colors, and fabric embellishments! We explain this in Section 2.

The mathematical goal of this chapter is to narrow down the quilt patterns to the so-called semiregular tessellations, as there are only finitely many of those, and we may have some hope of making a quilt from each of these patterns before our time on earth runs out. This narrowing down will comprise the bulk of Section 2, where we discover that there are no more than 12 of the desired quilt patterns. In Section 2.5, we briefly discuss a slightly more general class of quilt patterns about which very little is known.

Teachers of all levels will find useable ideas in Section 3. Quilting concepts are used in elementary education more often than most other fiber arts concepts. In this chapter we present some of those, but some more advanced concepts arise as well, ranging from approximations of reals by rationals, to continued fractions, to undergraduate research in *k*-uniform tessellations.

The sewing goal of this chapter is to give shortcut sewing instructions for most of the semiregular tesselation quilt patterns, and this will comprise the bulk of Section 4. The advantage of the shortcut techniques is not just in saved time but also in increased accuracy. Namely, fabric is stretchable, and the more one handles it by sewing and ironing it, the more distortion can occur in its size (especially along the edges, where it matters most) and the resulting quilt may not lie flat. With the methods described in this chapter, the pieces are often not cut until a later stage. This means that there are fewer edges to distort, or that the patterns are broken up into perhaps unusual but stabler shapes. Hence, less distortion occurs in the intermediate stages. A disadvantage of some of the presented techniques is that they use more fabric, some for the tucks and some for leftovers for another quilt top. (For years I have been making "left-over" quilts: they allow for greater freedom in placing pieces and colors and can be more pleasing than the more laboriously produced original quilts.)

Some of the described techniques are known among quilters, but most of them are probably not. They are presented here in the order of increasing difficulty. Readers are encouraged to experiment with these techniques, and to begin by trying the bed-sized squares-and-octagons quilt project in Section 4.2. Avid quilters should dive in head first (after reading the introductory sections about specialized techniques).

2 Mathematics

The aim of this section is to review the basics of tessellations, and to systematically pare down the infinite list of tessellating choices to a finite list. Much more on the subject of tessellations can be found in Grünbaum and Shephard [4], especially in Chapter 2.

2.1 Tessellations by Regular Polygons

A *tiling* or a *tessellation* of the plane is a countable collection of closed sets such that each point of the plane is in one of the closed sets, but that no point is in the interior of any two closed sets. A point may, of course, lie on edges of two different closed sets.

For our tessellations, all closed sets will be convex regular polygons. We also require that the intersection of any two distinct sets, if not empty, is either a single point or an edge of each of the two shapes. Such tessellations are called *edge-to-edge*. (This is in contrast to tessellations in which the intersection of two closed sets might be a partial edge, such as when in a square grid we shift alternate rows over by half the edge-distance.)

How many different regular polygon configurations can occur at a vertex of an edge-to-edge tessellation? If an n_1-gon, n_2-gon, ..., n_r-gon, meet at this vertex, the interior angles of the polygons have to add up to a full cir-

cle because there are no overlaps in the polygons. This is expressed succinctly in Theorem 2, which first needs the following lemma:

Lemma 1 *A regular polygon with n sides has interior angles measuring $\pi \frac{n-2}{n}$ radians (or $180 - \frac{360}{n}$ degrees).*

Proof Connect the vertices of the polygon to the center of the polygon. This produces n congruent isoceles triangles, each of which has its two equal sides incident at the center of the polygon. Each central angle is $\frac{2\pi}{n}$ radians. The other angles in the triangles, labeled α in Figure 2, have measure exactly one half of the interior angle of the polygon. In other words, the sum of the two non-central angles in each triangle equals the interior angle of the polygon. Since the sum of all the angles in a triangle is π radians, we conclude that the interior angle of a regular polygon with n sides equals $\pi - \frac{2\pi}{n} = \pi \frac{n-2}{n}$ radians. □

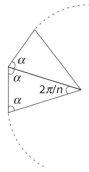

Figure 2. The interior angle of a polygon has measure twice α.

From Lemma 1, Theorem 2 follows directly: add the angles around a vertex and divide by π.

Theorem 2 *A regular n_1-gon, n_2-gon, ..., and an n_r-gon meet at a vertex without overlaps and without gaps if and only if*

$$\frac{n_1 - 2}{n_1} + \frac{n_2 - 2}{n_2} + \cdots + \frac{n_r - 2}{n_r} = 2.$$

We say that a point in a tessellation has *vertex configuration* $n_1, ..., n_r$ if regular n_1-, ..., n_r-gons meet with their vertices at the point without any overlaps and gaps (and the order in which the vertices meet is irrelevant). In the next subsection we determine all possible vertex configurations (see Theorem 3), and subsequently we examine when each vertex configuration can be extended to a tessellation of the plane.

2.2 Possible Vertex Configurations $n_1, \dots, n_r$

The goal is to find all integers $r \geqslant 3$ and $n_1, n_2, \dots, n_r \geqslant 3$ that are solutions to the equation $\frac{n_1-2}{n_1} + \frac{n_2-2}{n_2} + \cdots + \frac{n_r-2}{n_r} = 2$ from Theorem 2. It turns out that there are only finitely many solutions, and we list them below in Theorem 3. The list, of course, contains all the *regular* tessellations, namely tessellations using only one regular polygon: the vertex configurations are $3, 3, 3, 3, 3, 3$ for equilateral triangles, $4, 4, 4, 4$ for squares, and $6, 6, 6$ for hexagons. There are no regular tessellations using any other regular polygons.

Theorem 3 *The following is a complete list of integer solutions of the equation* $\frac{n_1-2}{n_1} + \frac{n_2-2}{n_2} + \cdots + \frac{n_r-2}{n_r} = 2$, *with* $n_1, \dots, n_r \geqslant 3$ *and* $r \geqslant 3$, *written in lexicographic (dictionary) order:*

(a)	$3, 3, 3, 3, 3, 3$	(j)	$3, 10, 15$
(b)	$3, 3, 3, 3, 6$	(k)	$3, 12, 12$
(c)	$3, 3, 3, 4, 4$	(l)	$4, 4, 4, 4$
(d)	$3, 3, 4, 12$	(m)	$4, 5, 20$
(e)	$3, 3, 6, 6$	(n)	$4, 6, 12$
(f)	$3, 4, 4, 6$	(o)	$4, 8, 8$
(g)	$3, 7, 42$	(p)	$5, 5, 10$
(h)	$3, 8, 24$	(q)	$6, 6, 6$
(i)	$3, 9, 18$		

The proof is technical, but straightforward. It proceeds by a case-by-case analysis, using lower bounds on the smallest n_i to obtain upper bounds on r, and then using the upper bounds to determine the possible n_i.

The list in the theorem gives all the solutions to a vertex configuration of an edge-to-edge tessellation using only regular polygons when we look at one vertex at a time.

2.3 Tessellating the Whole Plane

In this section we determine which vertex configurations from Theorem 3 extend to tessellations of the plane.

Consider strips made up of regular 3- and 4-gons as in Figure 3.

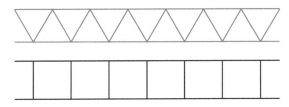

Figure 3. Strips of triangles and strips of squares.

The sides of the triangles and of the squares have equal lengths, so these strips can be pasted together in any order by aligning the vertices to get edge-to-edge tessellations. If we align only strips of triangles, we get the constant vertex configuration $3, 3, 3, 3, 3, 3$; if we align only strips of squares, we get the constant vertex configuration $4, 4, 4, 4$; if we alternate rows of triangles and of squares, we get the constant vertex configuration $3, 3, 3, 4, 4$; but there are also uncountably many other ways to arrange the strips (in correspondence with binary representations of real numbers in the unit interval).

Partially in the interest of being able to make a series of quilts in a finite amount of time, we now restrict our attention to tessellations with the same set of polygons at each vertex. (This prohibits arbitrarily mixing strips of squares and triangles.)

But even with this restriction, there are uncountably many quilts to make from regular polygons in an edge-to-edge manner. Namely, consider the strip of equilateral triangles and hexagons as in Figure 4.

Figure 4. A strip of hexagons, with corner triangles.

Such strips can be stacked in two different ways that have the same vertex configuration. See Figure 5.

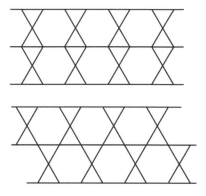

Figure 5. Two stackings of a strip of hexagons.

Thus even with this restriction we still get uncountably many distinct tessellations. How can we eliminate the uncountability of options?

We make one final restriction on the type of tessellations: we require that not only do the same polygons appear at each vertex, but that these polygons appear in the same order when enumerated either clockwise or counterclockwise. We call such tessellations *semiregular*.

We establish new notation for this purpose: if the regular polygons appear in the order n_1-gon through n_r-gon, either clockwise or counterclockwise around a vertex, that vertex has the *cyclic vertex configuration* $n_1.n_2.\ldots.n_r$. For example, in the tessellation at the top in Figure 5, the cyclic vertex configuration of a vertex on the central horizontal line is 3.3.6.6, 3.6.6.3, 6.6.3.3, or 6.3.3.6, and for a vertex between two horizontal lines it is 3.6.3.6 or 6.3.6.3. It is standard to record the configurations with the r-tuple that is the smallest in the lexicographic ordering. For the two vertices above, one would

thus record 3.3.6.6 and 3.6.3.6, respectively. Thus, a tessellation is semiregular if and only if all vertices have the same cyclic vertex configuration.

2.4 Determining All Semiregular Tessellations

We just defined a tessellation to be semiregular if the closed sets in the tessellation are all regular polygons, the polygons meet edge-to-edge, and at all vertices the cyclic vertex configurations are the same. Not all of the 17 vertex configurations on the list in Theorem 3 can be made into semiregular tessellations. In this section we review the list and determine which extend to one or more semiregular tessellations.

Clearly (a) 3, 3, 3, 3, 3, 3 makes the regular tessellation 3.3.3.3.3.3 by equilateral triangles (and there is no choice for how the triangles are joined). See the first tessellation in Figure 14.

Next on the list in Theorem 3 is 3, 3, 3, 3, 6, so the hexagon has to be surrounded by triangles, as in Figure 6.

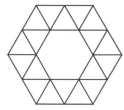

Figure 6. A hexagon surrounded by triangles.

Each of the triangles on the outside has to share a vertex with another hexagon. Thus, there are two distinct ways of continuing the construction with a hexagon at the rightmost vertex, see Figure 7. Once the placement of the second hexagon is chosen, the rest of the tessel-

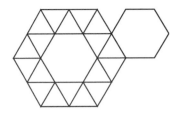

Figure 7. Two possible continuations of the hexagon-triangle combination.

lation is uniquely determined: in Figure 8 there is an extension of the first continuation. The extension of the other 3.3.3.3.6 continuation produces a reflection of this one; the two tessellations are not identical, but are isomorphic up to reflection.

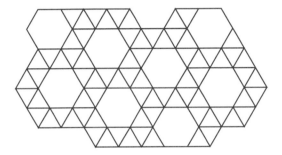

Figure 8. A 3.3.3.3.6 semiregular tessellation.

The next vertex configuration on the list in Theorem 3 is 3, 3, 3, 4, 4. This translates to possible cyclic configurations 3.3.3.4.4 and 3.3.4.3.4, and each gives a uniquely determined semiregular tessellation. See the bottom two leftmost tessellations in Figure 14. As discussed earlier, the 3.3.3.4.4 tessellation exists and is uniquely determined by alternating square and triangle strips. We need to verify that the 3.3.4.3.4 tessellation can be constructed in only one way, and that it is given by the tessellation on the bottom left in Figure 14 (in particular, that the partially depicted tessellation can be extended to the whole plane). We start from scratch: because the vertex configuration is 3.3.4.3.4, somewhere in the tessellation there are two equilateral triangles sharing an edge, as seen in Figure 9.

Figure 9. Building 3.3.4.3.4—start with two adjacent triangles.

Since no vertex has three triangles incident, these two triangles must be surrounded by squares, as in Figure 10. No squares are adjacent, so we have no choice also in surrounding these last squares by triangles. Continue this process; it produces the black lines in Figure 11. How can we be sure that eventually the entire plane will be tiled with the semiregular tessellation 3.3.4.3.4? Observe that the regions marked in red are copies of the fundamental domain, and that a 90° rotated copy of a red square precisely fills in the marked blue square. Thus if this part is forced and possible, so is the rest of the plane tessellation.

Figure 10. Building 3.3.4.3.4—squares surround the two starting triangles.

We proceed to the vertex configuration 3, 3, 4, 12. Abstractly, the cyclic vertex configurations could be 3.3.4.12 or 3.4.3.12. We first attempt to produce a 3.3.4.12.

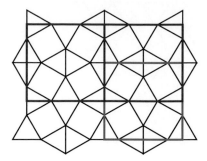

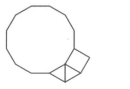

Figure 12. Attempting to build 3.3.4.12.

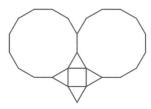

Figure 13. Attempting to build 3.4.3.12.

Figure 11. In 3.3.4.3.4, two red squares and one blue square are marked above; the center of any red-square edge could have been the start of the construction as in Figure 9. Rotate the blue square by 90°, translate, and obtain either of the red squares.

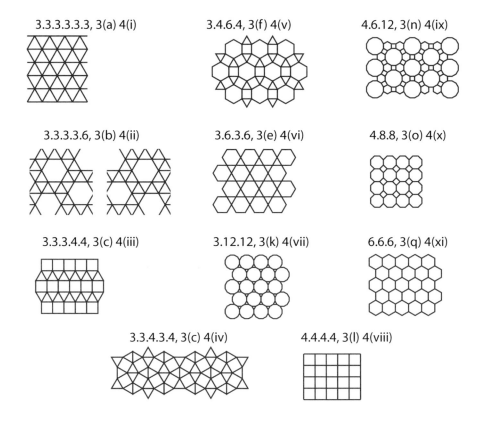

3.3.3.3.3.3, 3(a) 4(i)

3.4.6.4, 3(f) 4(v)

4.6.12, 3(n) 4(ix)

3.3.3.3.6, 3(b) 4(ii)

3.6.3.6, 3(e) 4(vi)

4.8.8, 3(o) 4(x)

3.3.3.4.4, 3(c) 4(iii)

3.12.12, 3(k) 4(vii)

6.6.6, 3(q) 4(xi)

3.3.4.3.4, 3(c) 4(iv)

4.4.4.4, 3(l) 4(viii)

Figure 14. All the possible semiregular tessellations: each tessellation is identified with its cyclic vertex configuration, its letter on the list in Theorem 3, and its roman numeral on the list in Theorem 4.

At each vertex there would be two adjacent triangles, and up to reflection, a square and a dodecagon must be adjacent at one of the vertices, as on the left in Figure 12.

Because triangles must be adjacent at each dodecagon vertex, this forces another triangle as on the right in Figure 12 in red color. This produces a 3.3.3 ... vertex, which is not allowed. Thus, 3.3.4.12 does not extend to a semiregular tessellation. Also, 3.4.3.12 does not produce a semiregular tessellation, as the squares share edges only with triangles, so a triangle-dodecagon-dodecagon vertex is forced as in Figure 13.

Next on the list is the vertex configuration 3, 3, 6, 6. One can show as above that 3.3.6.6 is not possible, as an attempt at construction forces some vertices to be 3.6.3.6. As we have seen (on the right in Figure 5), the semiregular tessellation 3.6.3.6 is possible.

The next on the list is 3, 4, 4, 6. There is no semiregular tessellation 3.4.4.6, which can be seen using an argument similar to that used to prohibit 3.4.3.12. However, there is a unique semiregular tessellation 3.4.6.4; see the first entry in the second column of Figure 14.

The reader can verify that there are no semiregular tessellations arising from configurations (g) 3, 7, 42, (h) 3, 8, 24, (i) 3, 9, 18, and (j) 3, 10, 15. Configurations (k) 3, 12, 12 and (l) 4, 4, 4, 4 give the unique semiregular tessellations depicted in the second column of Figure 14. The reader can straightforwardly eliminate (m) 4, 5, 20 from the list of semiregular tessellations, but (n) 4, 6, 12 and (o) 4, 8, 8 give the unique semiregular tessellations as the first two tessellations in the last column in Figure 14. Similarly, the reader can verify that (p) 5, 5, 10 produces no semiregular tessellation, but that (q) 6, 6, 6 gives the honeycomb semiregular tessellation, depicted in the last column of Figure 14.

We have thus (mostly) proved:

Theorem 4 *There are only 12 semiregular tessellations, up to translations and rotations, and there are only 11*

semiregular tessellations up to translations, rotations, and reflections (3.3.3.3.6 has a distinct mirror image):

(i)	3.3.3.3.3	*(vii)*	3.12.12
(ii)	3.3.3.3.6	*(viii)*	4.4.4.4
(iii)	3.3.3.4.4	*(ix)*	4.6.12
(iv)	3.3.4.3.4	*(x)*	4.8.8
(v)	3.4.6.4	*(xi)*	6.6.6
(vi)	3.6.3.6		

All this is illustrated in Figure 14. Note that in the semiregular tessellation 4.6.12, if counting clockwise, some vertices have a square (4-gon) followed by a hexagon followed by a dodecagon, and some have this order if counting counterclockwise. For all other semiregular tessellations, the orientation of counting is irrelevant.

2.5   *k*-Uniform Tessellations

For any two vertices in a given semiregular tessellation, it is possible to send one to the other while isometrically mapping the remainder of the tessellation onto itself. More generally we have the following definition:

Definition 3 A tessellation is called *k-uniform* if the vertices of the tessellation can be divided into k nonempty disjoint sets $V_1, \dots, V_k$ such that for any two vertices v and w there exists a rigid motion of the plane carrying the tessellation to itself, and v to w, if and only if v and w are in the same V_i. (A more technical way of saying this is that the symmetry group of the tessellation has exactly k transitivity classes of vertices.)

With this terminology, we observe that all semiregular tessellations are 1-uniform. Section 2.4 showed that there are exactly eleven 1-uniform edge-to-edge tessellations of the plane formed by regular polygons.

Grünbaum and Shephard [4] state that there are exactly twenty 2-uniform edge-to-edge tessellations of the plane formed by regular polygons. This was proved by Krötenheerdt [5] in 1969. Some readers may wish to carry out a proof of this case. Chavey [1] proved in the 1980s that there are exactly sixty-one 3-uniform tessellations. More recently, Galebach [2] wrote a program to compute the number of k-uniform edge-to-edge regular-polygon tessellations; the program computes the number of 4-uniform tessellations to be 151, the number for 5-uniform to be 332, and the number for 6-uniform to be 673 (after a month of computation). But there yet appears to be no *proof* of these results, and no further numbers are known as of this writing. The good news for a quilter is that for each k, the number of such k-uniform tessellations is finite—this occurs because there are only finitely many possible symmetry groups—so that only finitely many quilts have to be made for each k. The not so good news is that the number of quilts to be made is still very large for one's lifetime.

3 Teaching Ideas

Semiregular tessellations can be studied in a classroom at several levels. Building tessellations both combinatorially and geometrically is highly illuminating. The explorations in Sections 3.1 and 3.2 can be done exhaustively by elementary-school students or in a deeper way by more advanced students; some of the aspects of these investigations will even be instructive to graduate students. The questions in Section 3.3 ask the reader to determine the finished size of the n-gons on a quilt top given the cut size of fabric n-gons, and, conversely, the size n-gons one needs to cut in order to wind up with a specific size n-gon on the quilt top. The results are used in in the planning of the quilts in Section 4. The Pythagorean theorem and trigonometry are needed to successfully complete the Section 3.3 investigations. Section 3.4 uses continued fractions to find rational approximations to irrational fabric proportions. It requires fairly good algebra skills but no concepts encountered after high school. The research questions in Section 3.5 require excellent geometric and organizational skills.

3.1 Exploring the Tessellations Geometrically

The goal of the quilting constructions presented in Section 4 is to give ways to create semiregular tessellations efficiently and to minimize the propagation of error. After all, when tessellating the plane, accuracy of the produced regular polygons (most with irrational heights) is paramount, as is aligning edges and vertices precisely. The same issues arise when attempting to construct the tessellations on paper or on a computer. Here are some exercises designed to explore building tessellations. For students using paper, magnets, or feltboard, stacks of regular n-gons are needed, at least one for each of the various values of n. Suggested computer software includes Geogebra, Geometer's Sketchpad, or even Adobe Illustrator; each has its own challenges.

For each semiregular tessellation a student tries to create, the following problems can be posed.

* In the semiregular tessellations, the only regular n-gons used are 3-, 4-, 6-, 8-, and 12-gons. Construct each of these shapes with a straightedge and compass.

* One way of drawing a semiregular tessellation consisting only of squares is to draw the squares one at a time and then align the edges, but there is obviously a faster (and more precise!) way of drawing this semiregular tessellation—what is it?

* Which semiregular tessellations can be drawn more efficiently than one shape at a time (apart from the 4.4.4.4 tessellation mentioned above)? Discuss possibilities for shortcuts. Are there some parallel lines in the design? Can some lines or edges be extended?

* If you are using a computer program for drawing, what is the best way to replicate a diagram? Should you make one line segment at a time? Should you make all parallel line segments first? Is there some other set of line segments to start with, such as all the lines coming out of a vertex? How should these lines be made so that they are accurate (i.e., of the correct length and in the correct locations relative to each other)? Should you make regular polygons and then copy and paste those together? (How would one do that?)

* Suppose you were going to make a fancy floor (or a quilt) tiled with your tessellation. Try coloring your tessellations with colors (or fabrics) so that no two shapes sharing a whole edge have the same color. Which of the semiregular tessellations can be colored in this way using exactly two colors? Which ones cannot be colored with two colors, but three colors suffice? Do any of them need four or even more colors? (There is a big theorem in mathematics saying that all tessellations with connected pieces, semiregular or not, can be colored using no more than four colors. See [6].)

3.2   Exploring Nontessellations

The following two problems encourage students to fill in the details of eliminating vertex configurations that do not extend to tessellations of the plane.

* In Theorem 3, 17 potential semiregular vertex configurations are enumerated. According to Section 2, seven cyclic vertex configurations cannot be realized as semiregular tessellations. However, in Theorem 4, we found 11 different semiregular tessellations. Yet $17 - 7 < 11$. Explain what is going on here.

* The same drawing methods that were used for building the semiregular tessellations can also be used to help work through the proofs that some cyclic vertex configurations from Theorem 3 do not yield semiregular tessellations. Show that the following vertex configurations do not yield semiregular tessellations: (g) 3, 7, 42, (h) 3, 8, 24, (i) 3, 9, 18, (j) 3, 10, 15, (m) 4, 5, 20, and (p) 5, 5, 10. Note that two different cyclic vertex configurations may arise from (f) 3, 4, 4, 6. Show that exactly one is impossible.

3.3   The Size of the Shapes

Quilters and tessellaters have to make a lot of calculations while doing their craft: what final size is needed, how many basic units, how much of each fabric and color, etc. Much of that arithmetic is quite elementary, but it can quickly get into harder mathematics. Prerequisites for this section are the Pythagorean theorem and trigonometry and a willingness to get one's hands dirty. Depending on the audience, some teachers may want to break down the questions for their students.

Question 1 If regular n-gons are to be cut from a strip of fabric (or paper) of width h inches, what is the largest possible side length of the n-gon? For example, if $n = 3$, then h denotes the height of the triangle, and the answer is $\frac{2}{\sqrt{3}}h$. If $n = 4$, then certainly the answer is simply h. Answer the question for $n = 6, 8, 12$.

One can also reverse the question:

Question 2 If we want the side length of a regular n-gon to be d inches, what is the smallest width of a strip of fabric (or paper) from which we can cut the n-gon?

Quilters have to be more careful than that, however! It is not enough to cut a piece of fabric to the finished size; extra fabric is needed because seam thread must be sewn on the interior of the fabric in order for the seam to hold. For quilting, it is traditional to add a $\frac{1}{4}''$ seam allowance to all edges. For example, if the finished square is supposed to be $3''$ by $3''$, one needs to cut the square of 3.5″ by 3.5″ of fabric, as $\frac{1}{4}''$ will be eaten away by the seaming along each of the four sides.

The previous two questions can be rephrased for quilting purposes as follows:

Question 1q If regular n-gons are to be cut from a strip of fabric of width h inches, and if $\frac{1}{4}''$ is reserved along each edge of the n-gon for seam allowance, what is the largest possible finished side length of the n-gon? Do this for $n = 3, 4, 6, 8, 12$. (The answers might not be unique; see Figure 16.)

Question 2q If we want the finished side length of a regular n-gon to be d inches, what is the smallest width of a strip of fabric from which we can cut a polygon so that after subtracting $\frac{1}{4}''$ along each side for seam allowance from this polygon, we get the desired n-gon? Do this for $n = 3, 4, 6, 8, 12$.

For specific numerical values of h and d in the questions above, the "answers" may be provided by geometric construction instead of numerically.

3.4 Approximating Fabric Proportions via Continued Fractions

Quilting provides an excellent real-life application of continued fractions. The presentation below can be used within a number theory course, or in a recreational mathematics setting as early as middle school. In making quilts, one often has a predetermined finished size, depending on the size of the bed or on the size of the recipient. One also often starts with a predetermined idea of what the building blocks of the quilt will look like, and then the mathematical input is to calculate how many of those building blocks are needed to fill the desired finished size.

Specifically, suppose we want to tessellate a quilt with equilateral triangles, each of side length d inches. How should we choose the number of triangles per row, and number of rows in the quilt, so that the finished size is as desired (or close enough)? For example, how should p, the number of triangle side lengths in a row,

and q, the number of rows in the quilt, be chosen so that the finished quilt is square?

The height of the quilt will be $pd\frac{\sqrt{3}}{2}$ inches and the width will be qd inches. Because $\sqrt{3}$ is irrational, it is *not* possible to find positive integers p and q such that the finished quilt would be square. However, with fabric stretchability, one can get close: if (p, q) is $(7, 6)$ (see Figure 35), then $7 \cdot \frac{\sqrt{3}}{2} \cong 6.06218$, which is very close to 6, and a quilt with $p = 7$ and $q = 6$ *looks* square. Similarly, $15 \cdot \frac{\sqrt{3}}{2} \cong 12.9904$ so $(15, 13)$ is a good approximation, and $97 \cdot \frac{\sqrt{3}}{2} \cong 84.0045$, so $(97, 84)$ is even closer to square. But how were these pairs (p, q) derived? Trial and error are trumped by the systematic method introduced next.

Definition 4 A *continued fraction* is an expression such as

$$x = a_0 + \cfrac{1}{a_1 + \cfrac{1}{a_2 + \cfrac{1}{a_3 + \cfrac{1}{a_4 + \cfrac{1}{\ddots}}}}},$$

where a_0 is an integer, and all other a_i are nonnegative integers. If $a_n = 0$ for some $n > 0$, then all subsequent a_i are also 0.

Let's compute an example, say for $x = \frac{2}{\sqrt{3}} \cong 1.15470053837925$. (The motivation for this x is that for a square quilt we want the ratio $\frac{p\sqrt{3}}{2} : q$ to be (close to) 1, i.e., we want $p : q$ to be (close to) $\frac{2}{\sqrt{3}}$.) The largest integer smaller than or equal to x is 1. Thus we set $a_0 = 1$. So we will want to write $x = 1 + 1/something$, and solving gives $something = \frac{1}{(x-1)} \cong 6.46410161513775$. We then have to write $something$ as $a_1 + 1/something\ else$. The largest integer less than or equal to $something$ is 6, so we set $a_1 = 6$, and by solving we find

$$something\ else = \cfrac{1}{\frac{1}{x-1} - 6} \cong 2.15470053837925,$$

whence we set $a_2 = 2$. Note that the part to the right of the decimal point has appeared before!

Exercise The two quantities above *look* similar. Prove that indeed they are the same. Namely, prove that

$$\frac{2}{\sqrt{3}} - 1 = \frac{1}{\frac{1}{\frac{2}{\sqrt{3}}-1} - 6} - 2.$$

This justifies that all further a_{2n+1} are 6, and all a_{2n} are 2. One way to notate this continued fraction is as $\frac{2}{\sqrt{3}} = [1; 6, 2, 6, 2, 6, 2 \ldots]$. Just as we take decimal truncations of decimal expansions of real numbers, similarly we can take truncations of continued fractions.

1. The truncation $[1; 6] = 1 + \frac{1}{6}$ of $\frac{2}{\sqrt{3}} = [1; 6, 2, 6, 2, 6, 2 \ldots]$ gives $\frac{7}{6}$ (hence the pair $(7, 6)$).

2. The truncation $[1; 6, 2] = 1 + \frac{1}{6+\frac{1}{2}}$ of $[1; 6, 2, 6, 2, 6, 2 \ldots]$ gives $\frac{15}{13}$.

3. $[1; 6, 2, 6] = 1 + \frac{1}{6+\frac{1}{2+\frac{1}{6}}} = \frac{97}{84}$.

4. $[1; 6, 2, 6, 2] = \frac{209}{181}$, etc.

In general, the continued fraction of any real number x is obtained as follows: set a_0 to be the largest integer less than or equal to x; a_0 is the *floor* of x, denoted $\lfloor x \rfloor$. Notice that $x - a_0 \geqslant 0$. Let $a_1 = \lfloor 1/(x - a_0) \rfloor$. To continue, let $b_1 = 1/(x - a_0)$ and $b_n = 1/(b_{n-1} - a_{n-1})$. Then $a_n = \lfloor 1/(b_{n-1} - a_{n-1}) \rfloor$. If at any point a denominator is zero, stop—x is a rational number.

If $x = [a_0; a_1, a_2, a_3, \ldots]$ is a continued fraction expansion of a real number x, let a truncated continued fraction $[a_0; a_1, a_2, a_3, \ldots, a_n]$ be written as $\frac{p}{q}$ with integers p and q having no common factor. In a number theory class one may want to prove the following fact, but some may want to take it on faith: for any rational $\frac{r}{s}$ and $0 < s \leqslant q$,

$$\left| x - [a_0; a_1, a_2, a_3, \ldots, a_n] \right| \leqslant \left| x - \frac{r}{s} \right|.$$

In other words, of all the rational approximations, the truncated continued fractions are the best, as long as the denominators of the rational approximations do not exceed the denominator of the truncated continued fraction approximation.

Further Questions

* What if we wanted the finished quilt to be rectangular, with ratio 7 : 5 of height to width? What would be a good number q of rows with each row p triangle side lengths in width? Verify your answers numerically. (Hint: do a continued fraction expansion of $\frac{7 \cdot 2}{5\sqrt{3}}$.)

* What if we modify the layout of the quilt to have q rows of equilateral triangles, each $\frac{p}{2}$ triangle side lengths wide?

* How can continued fractions be used for tessellations with hexagons, or the other semiregular tessellations? (You will need the calculation results from Section 3.3.)

* Compute a few terms of the continued fraction of π, and compute its first few truncated continued fractions approximations. (Of course, you cannot get all the terms of the continued fractions, and since you have only a limited access to the real value of π, you cannot compute very many parts of the continued fraction.)

3.5   *k*-Uniform Tessellation Projects

Just as we derived from scratch all the possible 1-uniform tessellations, one could determine from scratch all the possible 2-uniform tessellations, and higher as well. A big part of this project is keeping track of the discoveries, especially when the number of possible tessellations gets beyond 20. Other possible searches might be for those *k*-regular tessellations that use only squares and triangles, or some other small combination of shapes.

Figure 15. A sampler quilt made from the various instructions in this chapter.

4 Crafting Semiregular Quilts

We now proceed to making semiregular tessellation quilts. A sampler is shown in Figure 15.

No regular n-gon in any tessellation is pieced; in other words, hexagons are hexagons, not pieced from triangles, and so on. The instructions below are of a general nature, in the sense that you pick the finished size d of the sides of the polygons, the number of rows and number of columns (or whatever the appropriate measurements are), and then by calculating you determine the dimensions of the cut pieces. Sample cut measurements are given in the cutting instructions for each tessellation, for experimentation purposes.

The presented techniques avoid having to sew any set-in seams; instead, when expedient, they create folds and tucks, and the finished product lies flat. (For this reason, we have omitted the semiregular tessellations 3.4.6.4 and 4.6.12, because the only known techniques either use set-in seams or are too complicated.) Primarily the presented shortcuts allow one to piece quilt tops faster and more accurately. (For the 6.6.6 tessellation, it is doubtful that the "quick" method provided here is actually quick.) Sometimes the shortcuts use more fabric than necessary, but the sewn remnants can be used for other quilting projects. There are no instructions for quilting and binding—for help with these techniques, see for example [3]. We recommend using a rotary cutter and quilter's ruler, and a triangle ruler makes it easier to accurately cut 30° and 60° angles.

Start by ironing the fabrics so that you can measure and cut precisely. (There are two opposing schools of thought on whether the fabrics should be prewashed; I always prewash.) Only rarely push the iron horizontally on the fabric—instead, lift the iron, press it down firmly, lift, etc.—horizontal glides stretch fabric in the direction of the glide and may distort the seam and edge, especially when the seam or the edge is cut on the bias (i.e., neither parallel nor perpendicular to the fabric selvage).

When pressing a seam, the crease should form as close to the seam as possible, not a millimeter away from it.

Remember to add seam allowance before cutting the fabric—seam allowance is the part that gets tucked away after forming a seam and will not be seen in the final product. The most common seam allowance is $\frac{1}{4}''$ or 6 mm. We explain polygon seam allowances here, though our instructions in the following sections account for seam allowances (thereby relieving stitchers of this duty). The explanation here is used implicitly in the later instructions, and may be needed by quilters who wish to use piecing techniques other than those given here (e.g., those who wish to attempt the 3.4.6.4 and 4.6.12 tessellations on their own). Such quilters may find English paper piecing [3, pp. 220–221] useful.

Mathematically, one can figure out exactly what size polygon pieces to cut to have the desired finished effect. However, even with precise sewing, the thread and the ironing eat up a bit of fabric, and some fabrics resist pressing more than others. Thus, the final product will deviate from the ideal; knowing your seam and iron allowances helps. (Note: quilters are used to measurements in increments of $\frac{1}{8}''$. The nature of working with regular polygons other than squares is that smaller increments (such as $\frac{1}{16}''$) are often needed as the relevant ratios may not be rational. Convert perplexing measurements (such as $d\sqrt{3}+\frac{1}{2}$ inches) to decimal form and then approximate closely to quantities measurable by an English ruler. For approximation assistance, see the sidebar on binary fractions on page 66 in Chapter 4.)

Observe that if more than one kind of polygon is used, the side lengths of the cut pieces differ, even though the side lengths of the finished ones will be the same! (Thus, when sewing two such edges together, first mark with pins on each fabric the two points that will become finished vertices. Sew the two edges together from pin to pin. Do not expect the two pieces to also align at the ends of the seam. With experience, you will be able to omit the pinning step.)

A Quick Lesson on Polygon Seam Allowance

This explains how the calculations were done for the tessellation quilting instructions. The diagram below shows part of a polygon with n equal sides; only the center and two sides are shown, with black denoting the finished edge and plum the edge with seam allowance (d denotes the finished side length and s the seam allowance).

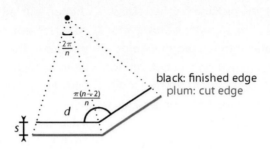

black: finished edge
plum: cut edge

Note that the assumption here is that the cut shape is a regular polygon of the same kind as the desired finished polygon.

The following chart may be helpful for constructing regular n-gon (n-sided polygon) pattern pieces with side length d and seam allowance s inches if $n \geqslant 4$:

cut side length $= d + 2s \tan \left(\frac{\pi}{n} \right)$

distance from center to cut vertex $= \dfrac{\left(d + 2s \tan \left(\frac{\pi}{n} \right) \right)}{2 \sin \left(\frac{\pi}{n} \right)}$

distance from center to cut edge $= \frac{d}{2} \cot \left(\frac{\pi}{n} \right) + s$

distance from center to finished vertex $= \dfrac{d}{2 \sin \left(\frac{\pi}{n} \right)}$

Thus, if $n = 6$ and $s = \frac{1}{4}''$, the cut side length is $d + 0.5 \tan(\frac{\pi}{6}) = d + \frac{1}{2\sqrt{3}} \cong d + 0.2887 \cong d + \frac{1}{4} + \frac{1}{32}$ inches. The distance from the center to cut side is $\frac{d\sqrt{3}}{2} + \frac{1}{4}$, so that if we cut hexagons from a rectangular strip, the strip should have width $d\sqrt{3} + \frac{1}{2}$ inches.

The chart above does not apply to triangles, as the angle at the center of the triangle is too large. The relevant information for triangles is that if we cut them from a rectangular strip, for cutting as on the left in Figure 16, the width of the strip should be $\frac{\sqrt{3}}{2}d + \frac{3}{4}$ inches, and for cutting as on the right in Figure 16, the width of the strip should be $\frac{\sqrt{3}}{2}d + \frac{1}{2}$ inches.

In the instructions above we assumed that the cut shape is the same type of polygon as the desired finished size polygon. It is possible, however, to cut differently. Two distinct cutting methods are illustrated in Figure 16 for equilateral triangles.

Figure 16. Two possible ways to cut a triangle with seam allowance.

If h is the altitude (height) of the finished triangle and the seam allowance is $\frac{1}{4}''$, one can cut as on the left from fabric strips of width $h + \frac{3}{4}$ inches (this is the same as the method above), and one can cut as on the right from fabric strips of width $h + \frac{1}{2}$ inches.

Well, it is time to stop talking and start quilting. The easiest quilt is the 4.4.4.4 tessellation given in Section 4.3—a beginner should start there. All quilters will appreciate this chapter's innovative quilting techniques. These techniques include

* tucks and temporarily unsewn seams, used in the tessellations 4.8.8 (Sections 4.1 and 4.2), 3.12.12 (Section 4.7), 3.3.4.3.4 (Section 4.9), and 6.6.6 (Section 4.10);

* sewing tubes at a slant, used in the second methods for the 3.3.3.3.3.3 (Section 4.4) and 3.6.3.6 (Section 4.6) tessellations.

Section 4.7 contains instructions on how to quickly cut hexagons from a strip of fabric.

In Sections 4.3–4.10, general instructions are given for tessellation quilts in approximately increasing order of difficulty. Instructions for some tessellations refer to instructions for previously described tessellations. Specific instructions for a bed-size 4.8.8 quilt are given in Section 4.2. To get to a quilt as soon as possible, we start with the 4.8.8 instructions in the first section below (Section 4.1). Happy sewing!

4.1 Construction of 4.8.8

My first attempt years ago at a 4.8.8 tessellation convinced me that piecing together individual squares and octagons is not for me—my attempted tessellation did not lie flat. This prompted the quicker and more accurate method given here.

Determine the number p of rows and the number q of columns of octagons; in Figure 17, $p = 3$ and $q = 9$.

Determine the finished length d of the sides of the octagons and squares: the total width of the quilt is then $q \cdot (1 + \sqrt{2})d$, and the height is $p \cdot (1 + \sqrt{2})d$.

Materials

* Two fabrics for octagons, each of total area $\frac{1}{2} \cdot p \cdot ((1 + \sqrt{2})d + \frac{1}{2}) \times q \cdot ((1 + \sqrt{2})d + \frac{1}{2})$ square inches, and

* A contrasting fabric for squares, of total area $(p + 1) \cdot (\sqrt{2}d + \frac{1}{2}) \times (q + 1) \cdot (\sqrt{2}d + \frac{1}{2})$ square inches.

Instead of cutting square pieces and octagon pieces and seaming them together, all fabrics are cut into squares. For lack of better nomenclature, the squares that eventually form octagons are called *octagon squares*, and the squares that produce squares are called *square squares*. The octagon squares are sewn together while at the same time catching the square squares by their vertices. The square squares are thus securely in place, but their sides are not sewn down (see Figure 18). After all the octagon squares are sewn together and the construction is carefully pressed, the loose square edges can be topstitched immediately or during the quilting.

Instructions

1. Cut the two octagon fabrics into $\lceil \frac{pq}{2} \rceil$ and $\lfloor \frac{pq}{2} \rfloor$ squares respectively, of side length $(1 + \sqrt{2})d + \frac{1}{2}$ inches. Pay special attention to the partial bracket symbol $\lceil x \rceil$ in the previous sentence. It means to

Figure 17. A finite part of the 4.8.8 tessellation instantiated as a table runner.

round up to the nearest integer and is called the *ceiling* of *x*. Similarly, the partial bracket symbol $\lfloor x \rfloor$ is called the *floor* of *x* and rounds down to the nearest integer. Cut the square fabric into $(p+1)(q+1)$ squares of side length $\sqrt{2}d + \frac{1}{2}$ inches. For experimentation, cut octagon squares of side length $4\frac{3}{4}''$ and square squares of side length $3''$.

Figure 18. Note the loose edges of the squares.

The next few steps are harder to explain than to do. Remember, the idea is to sew the octagon squares together while catching the square squares by the vertices. The octagons along the top of the quilt are treated differently, so read the instructions to the end before embarking on the sewing!

2. To sew together two octagon squares in the top row do the following. Lay one octagon square with right side up. Fold two square fabric squares in half, right sides out, without making a crease, and pin each with the long raw edges aligned with the top and bottom edges, respectively, of the octagon square, with the right side raw edges aligned, as in Figure 19.

   Lay the other octagon square on top, with right sides of the octagon fabrics together and with edges aligned.

   Sew along the $\frac{1}{4}''$ seam allowance, over both folded squares, but starting and stopping about $\frac{1}{2}''$ from the ends, as in Figure 20.

3. Repeat the previous step to join all octagon squares in the top row only. (Do *not* repeat for other rows!)

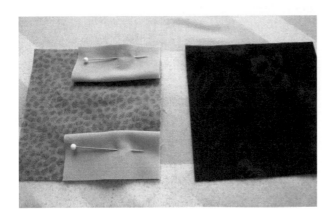

Figure 19. Preparation to sew together two octagon fabrics, with square squares to be caught in that seam.

Figure 20. Do not sew all the way to the edge; stop at least $\frac{3}{8}''$ away.

4. Presumably your quilt will have more than one row of octagons. How does one sew the octagons in the nth row to the previous rows? Consult Figure 21 (for simplicity, there is only one row above, and only one octagon pair in that row above).

Lay down the previous row(s), right side up. Grab two new octagon squares that you want sewn to the row above. Lay them below the previous row(s), right sides up, matching the colors correctly. In addition, pin one new square square in place (see Figure 21).

Figure 21. Setting up for sewing columns: there is one pinned yellow square on top of a red octagon square, and a loose blue octagon square ready for the red-blue octagon sandwich; above it is an already sewn blue-red octagon sandwich with two yellow folded squares.

Grab the loose end of the folded (yellow) square square above the two new octagons, and rotate it so that the raw long edge is aligned with the top edge of the unsewn red octagon square, aligning the raw right-side edges, as in Figure 22. The goal is to attach the lower square square from the top pair of octagon squares to the upper part of the bottom pair of octagon squares, so that it becomes the upper square square for the bottom pair of octagon squares. Pin in place, lay the right-side blue octagon square right-side down on top, aligning the right edges, and sew as in step 3. The result looks as in Figure 23.

5. Continue sewing to the previous rows, to get Figure 24. (If you are making a large quilt, you may want to get to the stage in Figure 24 more efficiently as follows (so you will not be constantly shifting between different jobs): first pin and sew the two leftmost octagon squares in the top row, as in step 2. Then, pin the third octagon square in the top row to the second square (as in step 2), and the second octagon square in the second row to the first (as in step 4), then sew the two seams (partially), one after

Figure 22. Preparation for sewing a row to a previous row: pin the folded square square from the row above to the octagon squares below—if you can ignore the existence of the row above, you really are doing exactly what was in step 2.

Figure 23. A part of a sewn column.

the other. Then pin the fourth octagon square in the first row to the third, third octagon square in the second row to the second, and second octagon square in the third row to the first, and sew the three seams partially. Repeat similarly, each time adding just one more octagon seam to each row. If you simply pinned the whole row to a previous row, you would have to reach for the scissors to clip threads to reposition one seam after the other, but with this "diagonal" method, you can sew all the pinned sandwiches one after the other without having to clip threads. I am all for time saving!)

Figure 24. All columns sewn together.

6. Next we sew the rows together. The square squares were originally folded down one middle, but we could have folded them down the other middle, and the idea is to perform that other fold now. Because parts of the square fabrics are already sewn down, it is harder to make this other fold, but it helps that in step 2 the seam stopped about $\frac{1}{2}''$ short of the edges. Be warned that this folding forces a 90° fold in place where eventually there will be a 45° fold, so there will be some bunching up in the rest of the

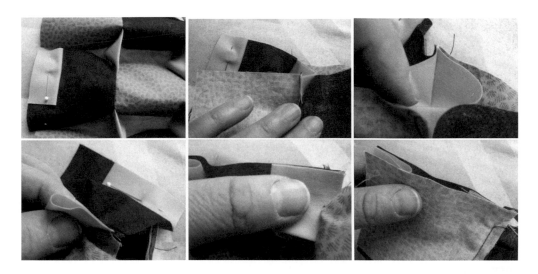

Figure 25. Sewing rows: two octagons at a time.

octagon while you are trying to keep the section to be sewn flat. Consult the photos in Figure 25.

In the upper left photo, a new folded square is pinned to the beginning of the row (as in steps 2 or 4). Then, the fun begins, folding differently the already folded (yellow) square squares. In the upper middle picture, the row of octagon squares on top was loosely folded over the row below. Note the pinned square from the first photo. Zoom in on the central (yellow) square square in the photo—that is the one that has to be refolded. In the upper right photo, the existing fold is opened; in the lower left photo, the other fold is made. It is aligned with the red octagon square in the lower middle photo.

Then it remains to lay the other octagon square on top, as in the lower right photo, then pin in place, and sew, partially, as before.

Repeat, to sew all the rows together. (As far as efficiency is concerned, I find it best to pin in this manner every second octagon pair in each horizontal row of desired seams, sew those, pin the rest of the pairs, and sew those.) The result is something like Figure 26.

Figure 26. All rows (and columns) are sewn together. Note that the octagon and square shapes are forming, but that the square edges are not sewn.

7. On the back side, clip the corners under the squares, just enough to prevent bulk, as in Figure 27. (Do not accidentally clip into the square squares on top that are supposed to hide these corners.)

Figure 27. Clip under the square centers.

8. Sew the remaining four folded square squares at the corners, as in Figure 28. These squares do not have to be sandwiched.

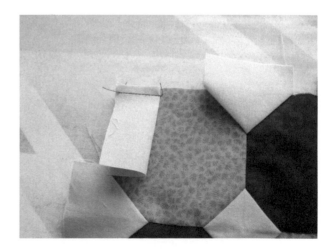

Figure 28. Sew the corner squares similarly.

9. Carefully press the seams. I find it best to first partially

press with the tip of the iron the octagon-octagon seam allowances *between* the square vertices from the back, without pressing on the squares, then turn the quilt top over. By possibly lifting the quilt a bit, adjust with your fingers all seam allowances under a square to lie flat, then press the square from the top. Having partially pressed the seam allowances before speeds up this process by a lot. The result is Figure 29.

10. Trim off the excess fabric. The result is ready for quilting. You may want to topstitch or quilt along the open square edges.

Figure 29. 4.8.8 tessellation: sewn, pressed, and ready for trimming.

Figure 30. A finite part of the 4.8.8 tessellation instantiated as a bed quilt.

4.2 Construction of a Bed-Size 4.8.8 Quilt

The instructions in this section are for a full bed-size quilt of size 76″ by 92″, as in Figure 30.

The quilt has 19 octagons in each of its 23 rows. (For symmetry of the trip-around-the-world pattern, you want odd numbers.)

Figure 31. After you cut a few squares (for the octagons and squares) you may still play around to determine which order of the fabrics is most pleasing. The photo shows a part of a quadrant of the finished pattern.

Each octagon square is cut with side length 4.5″, and each square square is cut with side length about 2.84″. You may want to think of 2.84 as about $2 + \frac{3}{4} + \frac{1}{16} + \frac{1}{32}$. One can get nine octagon squares out of a 42″ width of fabric, but I recommend not using the folded part of the bolted fabric (for fear of the fade on the fold),

in which case one only gets eight octagon squares per width. Similarly, one can get 13 or 14 square squares out of a 42″ fabric.

Choose a few fabrics that go together. I will give instructions as on the quilt in the picture, for nine fabrics. You may want to *audition* the fabrics, as in Figure 31, to determine the most pleasing arrangement.

Materials

There are $23 \cdot 19 = 437$ octagon squares and $24 \cdot 20 = 480$ square squares in the quilt.

⋆ Since 13 square squares can be cut off a 42″ wide bolt of fabric, you need to buy $\lceil \frac{480}{13} \rceil = 37$ square fabric widths of the fabric, namely, you need at least $37 \cdot 2.84 = 105.08″$ of fabric. Due to the inevitable shifts and slips that happen while cutting, you may want to buy somewhat more, say $3\frac{1}{8}$ yards for the square pieces. Choose a nondirectional fabric pattern, to save yourself extra effort.

⋆ There is a total of $19 \cdot 23 = 437$ octagon squares. One octagon lies in the center, surrounded by 4 of another color, followed by 8 of the next color, followed by 12 and so on. With nine fabrics, you need:

   – $1 + 36 + 12 = 49$ octagon squares of fabric in the center;

   – $4 + 38 + 8 = 50$ octagon squares of the second fabric;

   – $8 + 38 + 4 = 50$ octagon squares of the third fabric;

   – $12 + 36 = 48$ octagon squares of the fourth fabric;

   – $16 + 32 = 48$ octagon squares of the fifth fabric;

   – $20 + 28 = 48$ octagon squares of the sixth fabric;

   – $24 + 24 = 48$ octagon squares of the seventh fabric;

Figure 32. In each of the four quadrants, the columns—but not the rows—are sewn together. The central octagon and seams between the quadrants are still missing.

- 28 + 20 = 48 octagon squares of the eighth fabric;

- 32 + 16 = 48 octagon squares of the ninth fabric.

If eight octagon squares can be cut from the width of a 42″ bolt of fabric, you need at least $4.5 \cdot \lceil \frac{50}{8} \rceil = 4.5 \cdot 7 = 31.5″$ of the first three fabrics, and at least $4.5 \cdot 6 = 27″$ of the remaining six octagon fabrics. Buy a yard of each of the first three fabrics, and at least seven-eighths of a yard of each of the others.

Once you have the fabrics, cut them into squares, and follow steps 2–10 in the previous section. Here you have to keep track of the many colors, but the efficiency note in step 5 will help if you break the project into quadrants. Sew all seams around the inner squares in the four quadrants separately first, trim, press, then merge the four quadrants into one large quilt, and trim and press the result. See Figure 32 for two steps in the process.

4.3 Construction of 4.4.4.4

If you have never sewn before, this is the best project with which to start. We use only two fabrics, a dark one

and a light one, and the product is a checkered rectangle as in Figure 33.

Figure 33. A finite example of the 4.4.4.4 tessellation.

Determine the length d (in inches) of the finished squares, the number p of rows, and the number q of columns of the checkered grid. Of course, the finished product will then be pd inches by qd inches. An easy starting project is an 8×8 checkerboard of finished side length $d = 2″$.

Materials

* Two fabrics, a dark one and a light one, each of total area generously more than $\frac{p}{2} \cdot (d + 0.5)$ inches by $q \cdot (d+0.5)$ inches if p is even, and more than $p(d+0.5)$ inches by $\lceil \frac{q}{2} \rceil \cdot (d + 0.5)$ inches if p is odd. (Recall that the ceiling function $\lceil x \rceil$ rounds x up to the next integer.) Preferable dimensions of the two fabrics are $\frac{p}{2} \cdot (d + 0.5) + 2$ inches by $q \cdot (d + 0.5) + 2$ inches if p is even, and $p \cdot (d + 0.5) + 2$ inches by $\lceil \frac{q}{2} \rceil \cdot (d + 0.5) + 2)$ inches if p is odd. The extra $2''$ is for the inevitable shifts and slips that happen while cutting.

Instructions

1. Cut each of the fabrics into strips of width $d + 0.5$ inches. (For experimentation, this will be 2.5''.) If p is even, cut $\frac{p}{2}$ strips of each fabric (for a total of p strips), of length at least $q \cdot (d + 0.5)$ inches. If p is odd, cut p strips of each color (for a total of $2p$ strips), of length at least $\lceil \frac{q}{2} \rceil \cdot (d + 0.5)$ inches. The width should be precise, but the length can be compiled from several strips, keeping in mind that lengthwise you want to have q, respectively $\lceil \frac{q}{2} \rceil$, units of length $d+0.5$ inches. For greater precision (less distortion while handling), make the cuts either parallel or perpendicular to the fabric selvage.

2. Sew the strips, alternating between the dark and light color, into p-wide bands. If p is odd, half of the p-wide bands should have the dark color on the outside and the other half should have the light color on the outside.

3. On the sewn p-strip bands, press all seam allowances in the same direction (either all to the left or all to the right). If p is even and you have more than one band of strips, make sure that all the bands have identical pressing patterns.

4. Cut the bands into widths $d + 0.5$ inches, perpendicular to the seams, to get q checkered strips. (If p is odd, you will have one extra strip.)

Figure 34. Press seam allowances in opposite directions.

5. Sew the q new strips together into the desired checkered design. The seam allowances in the adjacent strips should go in opposite directions (see Figure 34). This makes for less fabric bulk at the corners (vertices), and the seams fall together naturally to create a more precise point. (You want the four squares to come together at exactly one point!)

6. Press the new seam allowances as desired. Proceed to quilting.

4.4   Construction of 3.3.3.3.3.3

Decide on the finished side length, d, or alternately, on the altitude $h = \frac{\sqrt{3}}{2}d$, of the finished triangles, and on the number of horizontal rows, p, and the number of triangle widths, q, in each row. The finished quilt will be of size $\frac{pd\sqrt{3}}{2}$ by qd. One option for achieving a desired finished size is to plan for a somewhat larger quilt and then trim it down in some aesthetically pleasing way, but—why give yourself more work in making more triangles?—the instructions below are for precisely p rows and precisely q triangle side-widths. Two examples of choices of p and q are in Figure 35.

The number of full triangles is $(2q-1)p$, but there are also $2p$ half-triangles. Because of seam allowances, the

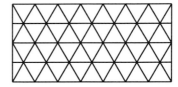

Figure 35. Two finite parts of the 3.3.3.3.3.3 tessellation: on the left $p = 6$, $q = 4$; and on the right, $p = 4$, $q = 7$.

2p half-triangles use up more fabric than p full triangles. For simplicity, we will treat all triangles, partial or not, as full triangles. Thus, in total there are $(2q + 1)p$ triangles.

Below are two sets of instructions for making the 3.3.3.3.3.3 quilts. The first set is easiest to follow, but the second method is much more time-efficient.

Materials

* Two fabrics, a dark one and a light one, each of total area generously more than $\lceil \frac{p}{2} \rceil \cdot (\frac{\sqrt{3}}{2}d + \frac{3}{4}) = \lceil \frac{p}{2} \rceil \cdot (h + \frac{3}{4})$ inches by $(q + 1) \cdot (d + \frac{\sqrt{3}}{2}) \cong (q + 1) \cdot (d + 0.866)$ inches.

* For the second method, you will need at least $\lceil \frac{p}{2} \rceil \cdot (\frac{\sqrt{3}}{2}d + 1.25) = \lceil \frac{p}{2} \rceil \cdot (h + 1.25)$ inches by $(q + 1) \cdot (d + \frac{1}{\sqrt{3}}) + \frac{\sqrt{3}}{2}d = (q+1) \cdot (d + \frac{1}{\sqrt{3}}) + h \cong (q+1) \cdot (d + 0.577) + h$ inches.

First Set of Instructions

This set of instructions is for a two-color triangle tessellation, and while it is conceptually simpler than the second set of instructions, it has the potential for more distortion, and many short seams are needed rather than fewer long ones. Recall that you need two pieces of fabric $\lceil \frac{p}{2} \rceil \cdot (\frac{\sqrt{3}}{2}d + \frac{3}{4})$ inches wide by $(q + 1) \cdot (d + \frac{\sqrt{3}}{2})$ inches long, one in each color.

1. On the back side of the light fabric, draw $\lceil \frac{p}{2} \rceil + 1$ parallel lines at distance $h + \frac{3}{4}$ inches apart, each of length

$(q + 1) \cdot (d + 0.866)$ inches (these lengths can be compiled from several fabric pieces; work on each individually). The first and the last lines can be arbitrarily close to the edges of the fabric. For experimentation, draw the lines 2.5" apart. (This corresponds to $d = 2$".)

On the back side of the light fabric, draw $\lceil \frac{p}{2} \rceil + 1$ parallel lines the length of the fabric. These are the solid horizontal lines shown in Figure 36. The first and the last lines can be arbitrarily close to the edges of the fabric.

2. Place the right sides of the dark and light fabrics together. For greater precision, you may want to iron the fabrics together without distorting the drawn lines, and pin the light/dark pair here and there.

3. Sew $\frac{1}{4}''$ from each side of each drawn line. For the first and the last lines, only sew on the side toward the middle. See Figure 36.

4. Without moving the parts, cut along the drawn lines (through all layers), as well as at 60° angles as illustrated in Figure 37. The two kinds of 60°-slanted lines intersect on the drawn lines.

5. For each of the cut triangle units, remove the unneeded stitches at the tip and unfold the longer seam

Figure 36. Solid lines are the drawn lines. Sew on the dashed lines on either side of the drawn lines, through the two fabrics.

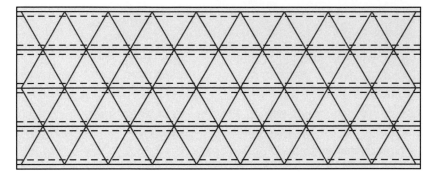

Figure 37. Dashed lines are the sewn lines; solid lines (red and black) are the cutting lines. (Here there are $\lceil \frac{p}{2} \rceil = 4$ rows and there are $q + 1 = 9$ full triangle widths in each row.)

to get a dark-light parallelogram as shown in Figure 38. You get $\lceil \frac{p}{2} \rceil (2q + 1)$ dark-light parallelogram units.

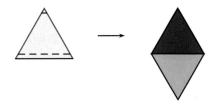

Figure 38. Open cut units from Figure 37.

(The forceful removal of the stitches somewhat damages the fabric at the tip; however, this is not a tragic disaster as first of all, very few stitches—and therefore weak stitches—are removed, and moreover, plenty of

fabric remains for the seam allowance. So, go ahead, use force to remove these stitches at the tip.)

6. Carefully, with a seam ripper, undo the longer seam stitches on $\lceil \frac{p}{2} \rceil$ of the dark-light parallelogram units. For the rest of the units, finger press the seams open, all the while making sure that you do not distort the seams or the sides. The sides are especially prone to distortion as they are cut on the bias (i.e., they are not cut parallel or perpendicular to the selvage).

7. Sew parallelograms together into rows. Sew q parallelograms per row; also sew to a color-appropriate end of each row *one* of the undone p triangles. In this way, you get p rows, each with q triangles of one color and $q + 1$ triangles of the other color.

8. The rows obtained in the previous step each contain $2q + 1$ triangles and are not rectangles. Carefully press the seams, making sure that you do not distort the size. Sew the rows together. For greater accuracy, I recommend that you pin the matched vertices first. Press again, trim to rectangle shape, proceed to quilting.

Second Set of Instructions

Compared to the first set of instructions, this one handles the triangles in a less fractured way. We rearrange sewn strips just as we did for the square grid method.

1. For finished triangle length d inches, cut $\lceil \frac{p}{2} \rceil$ strips of dark fabric and $\lceil \frac{p}{2} \rceil$ strips of light fabric, each of width $h + 1.25$ inches, where $h = \frac{\sqrt{3}}{2} d$ as before. For experimentation, cut the strips $3''$ wide (for $d = 2''$). The total length of the strips should be at least $(q + 1) \cdot (d + \frac{1}{\sqrt{3}}) + h$ inches.

2. Sew the strips together, alternating dark and light strips, but with the strips offset to form a $60°$ angle as in Figure 39. Eyeballing is fine. Otherwise, approximate this angle by folding a strip in half the short way, and subtracting the $\frac{1}{4}''$ seam allowance; use the remaining distance as the horizontal offset. For added precision, particularly when working with fewer or wider strips, you may want to start this step by cutting the tops of the strips at a $60°$ angle. Then simply line up the angled edges to get the effect of Figure 39.

Figure 39. Sew the strips with an offset, press, and cut at a $60°$ angle along the marked blue line.

3. Press all seam allowances in the same direction.

4. Make one cut across all the strips at precisely a $60°$ angle, close to the offset ends (along the blue line as in Figure 39). Discard the smaller part.

5. Sew the two outside strips together: line up the edges AB and CD as in Figure 40, but do not place A on top of C. A should be just to the left of C, as in the right picture in Figure 40. You obtain a tube as in Figure 41. The blue line in Figure 40 becomes the left rim of the tube in Figure 41.

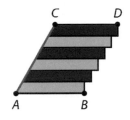

Figure 40. Align the lines AB and CD, so that the seam at $\frac{1}{4}''$ from the aligned edge hits AC. Beware: you can only overlay parts of the two lines AB and CD at a time, as a twist occurs.

Figure 41. The obtained tube.

6. Press the last seam in the same direction as all the other seams. This pressing is tricky as the seam is twisted around the tube.

7. Cut parallel to the smooth rim of the tube (line *AC*, through all layers), making $q + 1$ cuts that are $h + 0.5$ inches apart. This is $\frac{3}{4}''$ narrower than the width of the original strips. See photo in Figure 42.

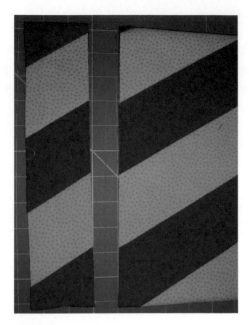

Figure 42. Cut the tube parallel to the rim, $h + 0.5$ inches apart (narrower than the width of the original strips).

8. The directions for sewing the $q+1$ tube strips together are lengthy and seem convoluted, but the process is really simple. Read through the remaining instructions before continuing to sew.

    Place one tube strip inside the other with right sides facing and approximately matching parallelograms of different colors. Make sure that the seam allowances on both tubes are oriented in the direction of sewing. To obtain an exact matching of seams that will leave room for new seam allowances, offset as in Figure 43: once the right edges of the top and the bottom tubes are aligned, the edge-wise distance between the seam on the top tube and the seam on the bottom tube is exactly twice the edge-wise length of the

seam allowance flap. Be very careful when sewing tubes together; the bias edges distort easily, so re-align at each intersection.

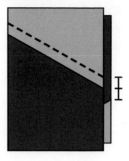

Figure 43. Align the two tubes vertically as shown—the two marked distances are equal. Before sewing, the right edges of the top and bottom tubes should be aligned.

9. Press the new seams either all open or all to one side. (Remember that these seams are on the bias, so you should press carefully to prevent distortion.)

Figure 44. Tubes sewn together with an offset.

10. The right side of the tube, at least a part of it, looks as in Figure 44. (It is easier to see the vertices on the right side of the fabric.) Mark one pencil line between the vertices as in the photo. This slanted line is theoretically exactly $\frac{3}{8}''$ from the vertices on either side, but in practice this distance will vary. This explains why the fabric strips in step 1 were cut $\frac{3}{4}''$ wider than the strips in step 7 (rather than only $\frac{1}{2}''$ wider).

The *idea* of the next few steps is to cut along the line marked in Figure 44, repeat the cuts along the parallel lines between other vertices, and then sew those strips together. *But* if we performed these cuts right now, the strips obtained by this approach would be parallelograms (instead of rectangles), so the end result would not be a rectangular quilt of the planned size. Thus a few additional steps are needed as given below.

11. Cut with scissors through one layer only, on the line marked in the previous step.

12. Join the the left and right edges from Figure 44, following the offset instructions as in step 8. This produces another twisted tube. Press the seam in the same way as you pressed the seams parallel to it.

13. We now open the tube to a single layer. Mark a line perpendicular to the cut edges of the tube (and to the marked line in Figure 44), steering away from any vertices by at least $\frac{1}{4}''$. Cut with scissors along this line, one layer only, to obtain a rectangle.

14. Cut between the vertices, as in Figure 44, parallel to the line in that photo.

15. Arrange the strips so that all triangles of the same color are oriented the same way, as in Figure 45. Lay the rows on a surface so that the left-right offsets of different rows form the largest possible rectangle. (If you made the cut in step 13 too close to the vertices, you still have to make the offsets, but your finished quilt width will have $q - 1$ triangles rather than q triangles.)

Figure 45. Cut sewn tubes between vertices.

Figure 46. Parts of a 3.3.3.4.4 semiregular tessellation.

16. Sew the strips together as arranged in the previous step. You may want to pin the matched vertices first, and adjust the triangle edges by feeling them through the layers. If the seam allowances on the top and bottom triangle edges go in opposite directions, "lock" the edges until snug, to enhance precision.

17. Press the obtained 3.3.3.3.3.3 tessellation, trim to desired size, and proceed to quilting and binding.

4.5 Construction of 3.3.3.4.4

This semiregular tessellation is constructed from (rows of) the 3.3.3.3.3.3 and 4.4.4.4 tessellations. Create such rows, making sure that the sides of desired finished triangles and squares are equal. (The suggested cut measurements for experimentation both have finished side length $d = 2''$, as above.) I recommend making the rows of triangles first, then measuring off the actual finished triangle side, and finally making the square strips with this measurement. The results are shown in Figure 46.

4.6 Construction of 3.6.3.6

A sample part of a 3.6.3.6 semiregular tessellation is shown in Figure 47.

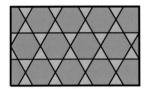

Figure 47. A finite part of the 3.6.3.6 tessellation.

The arrangement shows three horizontal rows of blue hexagons interspersed with light triangles; the instructions below first explain how to construct the rows.

Decide on the finished side length d of the hexagons and triangles, or alternately on the altitude $h = \frac{\sqrt{3}}{2}d$ of the triangles, on the number p of rows ($p = 3$

above), and on the total number q of hexagons or partial hexagons ($q = 15$ above; but if d is more than $1''$ and the hexagon fabric has a 180° rotational symmetry, 14 hexagons would suffice, as the top left and the center-row right partial hexagons can be made from one whole one).

Two methods are given below. The second one uses more fabric but fewer measurements. Each method leaves sewn remnants that are ready for another (non-semiregular tessellation) quilt.

Materials

Choose a fabric for the hexagons and a contrasting fabric for the triangles.

★ You need at least $q \cdot \frac{2d+1}{\sqrt{3}}$ inches by $2h + 0.5$ inches of the hexagon fabric.

★ For the triangle fabric, you need at least $q \cdot \frac{2d+1}{\sqrt{3}}$ inches by $2h + 1.5$ inches.

First Set of Instructions

1. Cut strips of the hexagon fabric of width $2h + 0.5$ inches wide. For the triangle fabric, cut strips of width $h + 0.75$. Experimenters can cut hexagon fabric 6″ wide and triangle fabric 3.5″ wide. (If you prefer to be more efficient with your use of fabric as in Figure 16 (right) on page 201, you may adapt this method to use strips of width $h + 0.5$ inches. Here experimenters can cut triangle fabric strips 3.25″ wide.) The total length of the hexagon fabric strips should be generously more than $q \cdot \frac{2d+1}{\sqrt{3}}$ inches, and the total length of triangle fabrics should be double that length.

2. Sew a strip of triangle fabric to each long side of every strip of hexagon fabric.

3. Press the seam allowances. (Any direction is fine.)

4. Now we will cut the strips into pieces as in Figure 48. Using a rotary cutter, make parallel 60° cuts every

2h+0.5 inches (same as the width of the hexagon fabric). Mark a point along one of the cuts at the center of the band width (shown in red in Figure 48). Cut through the mark at the other 60° angle, and make parallel cuts every 2h + 0.5 inches as in Figure 48.

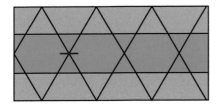

Figure 48. The slanted solid black lines are the cutting lines for 3.6.3.6. The red mark in the center of the strip shows the intersection of 60° cut lines.

Another way to get these parallelograms is to fold the sewn strips in half lengthwise and cut with a quilter's triangle ruler: the parallel lines on the ruler should align with the fold, while the tip of the triangle should be exactly on the edges.

5. Sew the parallelograms into rows as shown in Figure 49. Be sure to match adjacent parallelograms so that the triangles and hexagons meet at one point after sewing (i.e., match at the seam allowance). Because the edges are cut on the bias, handle the pieces carefully to avoid distortion.

Figure 49. Sew the obtained parallelograms into rows.

6. Convert the parallelogram rows into rectangles. For some rows, this will require cropping and for some rows additional triangles are needed. You may create those triangles from the spare triangle shapes created in step 5, by cutting along the red lines indicated in Figure 50 (or from uncut fabric). Press any new seams.

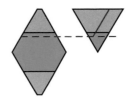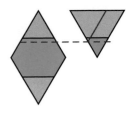

Figure 50. Additional triangles may be salvaged from the leftovers of step 4. Cut on the slanted solid red lines as shown, and then remove bulk by removing the remaining seam. Note that the height of the triangle is already correct.

7. Sew the rows together. (First pin at desired vertices for added precision.) Again, as the edges are cut on the bias, handle carefully so as not to get distortions.

8. Press the latest seams, and trim to desired size and shape.

Second Set of Instructions

1. Cut the hexagon fabrics into strips of width 2h + 0.5 inches and the triangle fabric into strips of width 2h + 1.5 inches. The number of strips of each fabric should be the same. Experimenters can cut hexagon fabric 6″ wide and triangle fabric 7″ wide.

2. Sew the strips together, alternating the two fabrics; offset at a 60° angle, just as for the 3.3.3.3.3.3 tessellation or as in Figure 51. Notice in Figure 52 that some stripes are slightly narrower (and the two outside strips look wider as not all seam allowances have been eaten away there by the seams).

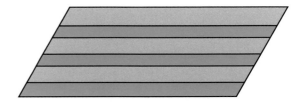

Figure 51. Sew the strips of hexagon and triangle fabrics at an offset.

3. Press the seam allowances all in one direction, and neaten *one* edge of the parallelogram by making a 60° cut along the edge of the strips. This resembles the second method for the 3.3.3.3.3.3 tessellation! Follow the instructions in Figure 40 to sew the two outer horizontal strips together. You obtain a tube. Press the new seam allowance in the same direction as all the other ones.

4. (This differs from the 3.3.3.3.3.3 tessellation.) Mark a line at 60° from the cut line but not parallel to the strips, as in Figure 52. It is best if the marked line meets the cut edge in the middle of a triangle fabric strip.

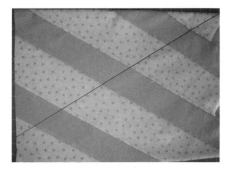

Figure 52. Mark a line at 60° from the cut line. (The photo shows the back side.)

5. Cut along this line with scissors, one layer only, to obtain the parallelogram shown in Figure 53.

Figure 53. The result of cutting the tube at the other 60° angle.

6. Cut strips parallel to the last cut, of width $2h + 0.5$ inches, but do not move the strips after cutting. Now cut at 60° parallel to another clean edge at $2h + 0.5$ inch intervals, meeting the first set of cuts in the middle of the fabric strips. See Figure 54.

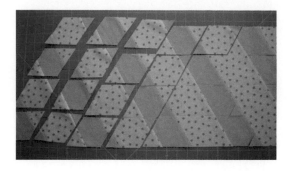

Figure 54. Cutting a larger parallelogram into smaller parallelograms. Some of the cut pieces were moved so you can see the cuts clearly.

Proceed as in steps 5–8 of the first method, using the parallelograms with the hexagon fabric (*not* the triangle fabric) in the middle. The remaining parallelograms are not used in this project, and in fact do not build a semiregular tessellation. The picture in Figure 55 shows (a part of) the tessellation together with a few sewn strips of remnants.

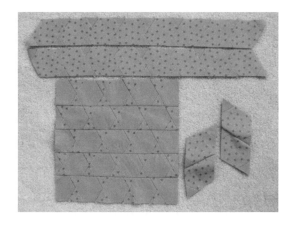

Figure 55. A 3.6.3.6 tessellation using the second method, with leftovers.

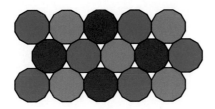
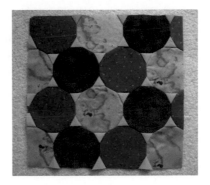

Figure 56. The 3.12.12 tessellation, drawn (left) and pieced (right).

4.7 Construction of 3.12.12

The method for the 3.12.12 tessellation is similar in spirit to the 4.8.8 construction: the dodecagon fabrics are sewn together while catching the triangle vertices, and at the end, the triangle edges are loose. You should understand the 4.8.8 construction from Section 4.1 before proceeding with this section.

Determine the number of dodecagons needed; refer to Figure 56, or sketch your own design.

Using three different colors for the dodecagons will assure that no two adjacent dodecagons have the same color. Use a different fabric for the triangles. Let d be the finished side length of the polygons. Choose d to be at least 1″, and if you have not had much experience sewing, make d at least 2″.

Instead of cutting dodecagons and triangles, you will be cutting hexagons. For lack of better nomenclature, the hexagons that eventually form dodecagons are called *dodecagon hexagons*, and the hexagons that

make triangles are called *triangle hexagons*. The dodecagon hexagons will be sewn together while catching the triangle vertices. All seams are partial—they start and end about $\frac{1}{2}''$ away from the edges.

Cutting Hexagons

Here are instructions on how to quickly cut hexagons whose opposite sides are x inches apart. The yield is not only hexagons, but also many triangles of side length $\frac{x}{\sqrt{3}}$ inches. Each hexagon takes x inches by $\frac{2x}{\sqrt{3}}$ inches of fabric. Here are two methods.

Cut strips of fabric of width x inches. Align the strips one on top of each other but at most six at a time. (If you are a beginner, start with one strip at a time.) Cut the obtained conglomerate strip at a 60° angle, and make parallel cuts every x inches. Mark a point along one of the cuts at the center of the strip, as in Figure 57. Cut through the point, making the other 60° angle, and make parallel cuts every x inches. Voilá, there are hexagons, with about twice as many triangles.

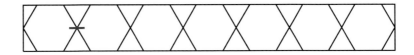

Figure 57. Cut along the black lines to quickly make many hexagons. Make sure that the angled lines meet in the center of the fabric strip.

The second method for cutting hexagons, suggested by Jeri Riggs, is to fold the fabric strip lengthwise in half, with the folded side down facing you. Align a 60° triangle ruler with point up so the distance along the folded edge is $\frac{x}{\sqrt{3}}$. Voilá, a finished hexagon with cut sides of length $\frac{x}{\sqrt{3}}$, and some left-over triangles.

Materials

⋆ For each dodecagon (of finished side length d inches), you need $\frac{d}{\tan(\pi/12)} + \frac{1}{2} \approx 3.73d + \frac{1}{2}$ inches by $\frac{2d}{\tan(\pi/12)\sqrt{3}} + \frac{1}{\sqrt{3}} \approx 4.31d + 0.577$ inches of fabric.

⋆ For each triangle, you need $d + \frac{1}{2}$ inches by $\frac{2d}{\sqrt{3}} + \frac{1}{\sqrt{3}} \approx 1.155d + 0.577$ inches of fabric.

Compute your total fabric requirements for each fabric, taking into account how many dodecagon and triangle pieces of each fabric you need.

Instructions

1. You may want to use the instructions above for cutting hexagons. For each dodecagon, cut a hexagonal fabric piece whose opposite sides are at distance $x = \frac{d}{\tan(\pi/12)} + \frac{1}{2} \approx 3.73d + \frac{1}{2}$ inches apart. For each triangle, cut a hexagonal fabric piece whose opposite sides are at distance $x = d + \frac{1}{2}$ inches apart. The sample shown in Figure 56 has finished size $d = 1$, with dodecagon fabric width $4.23 \approx 4\frac{1}{4}''$ and triangle fabric width $1.5''$. Beginning experimenters should use $d \geq 2$; for example, cutting dodecagon fabric width $9\frac{13}{16}''$ and triangle fabric width $3''$ corresponds to $d = 2.5''$.

2. To sew two dodecagons together while catching two triangle vertices, fold the two triangle hexagons in half through opposing vertices, right sides out and without forming a crease, and lay them down on one of the dodecagon hexagons. As in Figure 58, align the short triangle hexagon edges with the vertical raw dodecagon hexagon edges.

Figure 58. Preparing to sew two dodecagon hexagons together while catching the triangle hexagons in the seam.

3. Lay the other octagon on top of the construction in the previous step, right side down, aligning the vertices of the octagon hexagons. Sew along the seam allowance, but to *at most* $\frac{1}{2}''$ from the edge of the fabric.

4. Sew the top row of dodecagons in this way. Pay attention to the colors of successive dodecagons. See Figure 59.

Figure 59. A sewn row of dodecagons.

5. To sew further rows, you add one new dodecagon hexagon at a time, catching the already partially sewn

Figure 60. Adding a dodecagon hexagon to an already sewn unit of two dodecagon hexagons.

triangle hexagons differently, and adding a new triangle hexagon where needed. First, attach the red dodecagon hexagon to the already sewn unit of green and blue dodecagon hexagons. Start with the edge between the green and the red hexagons as shown in Figure 60 (left); a new folded triangle is pinned at the outside corner. The next step is to refold along another axis the triangle already caught between the green and the blue hexagons, then align the red-green hexagon edges, and sew that edge, starting and stopping at least $\frac{1}{2}''$ from the ends. The result is shown in Figure 60 (middle). That photo also shows yet another new pinned triangle. Figure 60 (right) shows the remaining blue-red edge ready to be sewn, but the blue hexagon fabric was pulled back to expose the folds underneath. Note that there are three ways of folding a hexagon through vertices; you have already made one fold, and to sew another dodecagon hexagon to this triangle hexagon piece, you will have to perform the other two folds, one for each of the two seams needed to add the third dodecagon piece. (The idea is really the same as the one in step 4 in Section 4.1 for refolding the square squares to attach octagon squares.)

6. Finish all remaining seams. Note that triangles are formed from the triangle hexagons; their vertices are

secured, but the edges are loose. You should get something like Figure 61.

Figure 61. Sewn 3.12.12 tessellation, ready for pressing.

7. From the back side, carefully trim excess bulk where the dodecagon hexagons come together. Make sure that you do not clip into the triangles above.

8. Carefully press each seam allowance and triangle. Trim the construction to desired size. Quilt.

4.8 Construction of 3.3.3.3.6

These instructions are for a three-color tessellation as shown in Figure 62: all hexagons are of the same color, and the triangles come in two colors. The shortcut for this pattern is to partition it into manageable sewing units, as in Figure 63. The whole tessellation consists of *large parallelograms* (each composed of two hexagons and twelve triangles), and small parallelograms, called *4-triangle units*. (Only four large parallelograms are labeled in the figure, but for planning purposes you should mark all of them on your own design.)

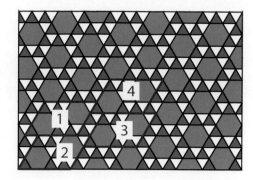

Figure 63. The 3.3.3.3.6 tessellation is broken into (equal) large parallelograms labelled 1, 2, 3, 4, and small (4-triangle) parallelograms between the large parallelograms.

Materials

From the two triangle fabrics, you will make c 4-triangle units, b units of 6-triangle rows, and $3b$ extra unsewn triangles of each color. Therefore, you will need to buy fabric:

* for the hexagons, $2h + \frac{1}{2}$ inches by at least $a(2d + \frac{1}{\sqrt{3}})$ inches;

* for the 6-triangle units, $\lceil \frac{b}{2} \rceil (h + 1.25)$ inches by $3(d + \frac{1}{\sqrt{3}}) + h$ inches;

* for the loose triangles, $h + 0.75$ inches by $\lceil \frac{2}{3}3b \rceil (d + \frac{1}{\sqrt{3}}) + d$ inches;

* for the 4-triangle units, $\lceil \frac{c}{2} \rceil (h + 1.25)$ inches by $2(d + \frac{1}{\sqrt{3}}) + h$ inches.

The following method is fast, but not fabric efficient; it is modified from the second method for the 3.3.3.3.3.3 tessellation in Section 4.4. Throughout, the fabric has many sides cut on the bias, so handle the pieces carefully at all stages.

Instructions

1. Cut the a hexagon pieces from strips of width $2h + \frac{1}{2}$ inches using the instructions on page 219. Cut the $3b$ loose triangle pieces in each triangle color from

Figure 62. A finished (but not quilted) 3.3.3.3.6 tessellation.

Choose the finished edge length d, or alternatively, the altitude $h = \frac{\sqrt{3}}{2}d$ of the triangles. Draw out the 3.3.3.3.6 semiregular tessellation and then impose a rectangle to get the design for your quilt. Count the number a of hexagons in your drawing, count the number b of the large parallelograms, and count the number c of 4-triangle units remaining. For simplicity, your counts should include even the partial units.

strips of width $h + \frac{3}{4}$ inches: an edge of each triangle is on one edge of the strip and its opposite vertex is on the other edge of the strip. Experimenters can cut hexagon fabric 6″ wide and triangle fabric 3.5″ wide.

2. Each large parallelogram unit contains two hexagons. The left hexagon has dark triangles attached to every other edge to form a large pointing-up triangle; the other hexagon has light-colored triangles attached to every other edge to form a large pointing-down triangle. Sew b of each of these large triangle units.

Figure 64. Broken down units for the 3.3.3.3.6 tessellation. Note the tilt of the four-triangle unit: only one tilt works!

3. Sew the c 4-triangle units, alternating dark and light colors, and all slanted in the same way as in the numbered units in Figure 63. See Figure 64. I recommend following steps 1–9 in the second method for the 3.3.3.3.3.3 tessellation with p equal to a and $q + 1$ in step 7 equal to 2 (this means that in step 7, two narrow tubes are obtained). Then draw the line down the middle of some slanted row of parallelograms, as in Figure 44. You have many choices for this line, and in fact, you have to cut through all of them eventually. It is easiest if you simply mark one of them, cut that

line with scissors, one layer only, and then proceed by making the other cuts with a rotary cutter.

4. Sew the b 6-triangle units, alternating dark and light colors, and all slanted in the same way as in the numbered units in Figure 63. *Important:* the slant of the 6-triangle units is different from the slant of the 4-triangle units! (See Figure 64.) I recommend following steps 1–9 in the second method for the 3.3.3.3.3.3 tessellation with p equal to b, and $q + 1$ in step 7 equal to 3, but with the following *important* modification starting in step 2: the 60° angle should slant upward and to the left instead of upward and to the right! When you are finished with steps 1–9, draw a line as in Figure 44, but of course slanted the other way. Cut that line with scissors, one layer only, and then cut the remaining parallel cuts down the centers of slanted rows of parallelograms. You have now completed all the basic parts of the tessellation.

5. Complete the b large parallelogram units: sew two large triangles (with tips of different colors) to a row of six triangles, as in Figure 64.

6. It remains to sew the b large parallelograms to the c 4-triangle parallelograms. Observe in Figure 63 that no adjacent units meet edge-to-edge. Thus, to have the quilt lie flat and also avoid set-in seams, we will make partial seams at each stage that are finished in later stages.

For example, referring to Figure 63, first sew the right edge of large parallelogram 1 to the 4-triangle unit, starting from the bottom edge and finishing about 1″ before the end of the 4-triangle unit. Then sew the top edge of parallelogram 2 to the bottom of the just-constructed unit, starting at the right and finishing about 1″ before the end of parallelogram 2. Next sew the left edge of parallelogram 3 to what you have just sewn, starting from the top edge and finishing about 1″ before the end of parallelogram 3. Then add parallelogram 4, starting at the left end and finishing before the end of parallelogram 4. At this point, the 4-triangle unit is almost sewn in, only the seam connecting with parallelogram 1 is not yet finished. You *could* finish that seam off right now, to complete sewing the 4-triangle unit in the center, but it would be simpler to finish it as part of a longer seam. For example, you could take the 4-triangle unit to the left of parallelogram 4 and sew it partially to parallelogram 1, and finish by sewing from the end of the first partial seam on the old 4-triangle unit to the top of parallelogram 4. Do you see the how partial seams become complete? You will quickly get the hang of what should be done next.

Keep adding the other large parallelograms and 4-triangle units, making some seams partial and completing some other seams, until everything is sewn together completely.

7. When all seams are completed, press, trim to desired size, and quilt.

4.9 Construction of 3.3.4.3.4

One could create a 3.3.4.3.4 quilt by imposing a grid on the tessellation as in Figure 65, piecing half-triangles to squares and thus making larger squares that are easy to assemble.

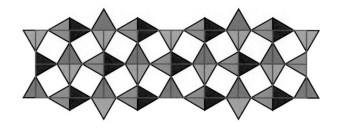

Figure 65. An impostor! Imposing a grid on the 3.3.4.3.4 tessellation.

Figure 66. A finished 3.3.4.3.4 top, with the square-triangle edges securely pressed in place, but not (yet) sewn.

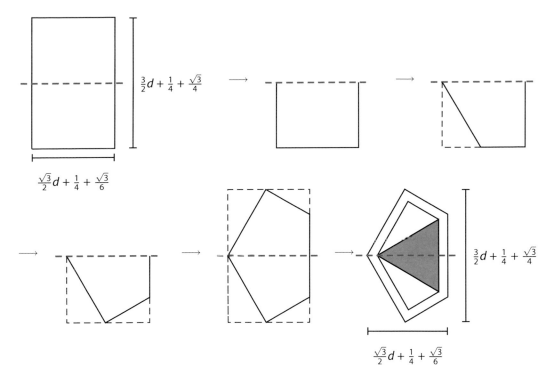

$\frac{3}{2}d + \frac{1}{4} + \frac{\sqrt{3}}{4}$

$\frac{\sqrt{3}}{2}d + \frac{1}{4} + \frac{\sqrt{3}}{6}$

$\frac{3}{2}d + \frac{1}{4} + \frac{\sqrt{3}}{4}$

$\frac{\sqrt{3}}{2}d + \frac{1}{4} + \frac{\sqrt{3}}{6}$

Figure 67. How to cut the triangle shapes for the 3.3.4.3.4 semiregular tessellation.

One might call this a *mock* 3.3.4.3.4 tessellation because each equilateral triangle would be composed of two pieces of fabric. But this seems like an improper way to achieve the semiregular tessellation 3.3.4.3.4, so we give a different method here, and a final product is shown in Figure 66. Here we sew squares together while catching the vertices of the triangles, just as in the 4.8.8 and 3.12.12 tessellations. Decide on the single finished side length d of both the triangles and the squares in the tessellation. (Consider choosing d so that $1.366d + 0.5$, which is the cut length of the triangle sides, is a number close to a multiple of $\frac{1}{8}''$. Trigonometrically, this is $d(\sin(\frac{\pi}{6}) + \cos(\frac{\pi}{6})) + 0.5$.) Use a schematic of your finished quilt to determine the number of pieces needed from each fabric. Say you need a squares and b of each of the triangle fabrics.

Materials

* You need a times $1.366d + 0.5$ inches by $1.366d + 0.5$ inches of the square fabric.

* Of each of the triangle fabrics, you need $\frac{\sqrt{3}}{2}d + \frac{1}{4} + \frac{\sqrt{3}}{6}$ inches by b times $\frac{3}{2}d + \frac{1}{4} + \frac{\sqrt{3}}{4}$ inches.

Instructions

1. For each finished square, cut a square of side length $1.366d + 0.5$ inches. Experimenters may use 3.5″ for smaller squares, or 6″ for larger squares.

2. Now cut rectangles that will result in triangles. Cut a strip of fabric of width $\frac{\sqrt{3}}{2}d + \frac{1}{4} + \frac{\sqrt{3}}{6}$ inches, and then cut the strip into rectangles of length $\frac{3}{2}d + \frac{1}{4} + \frac{\sqrt{3}}{4}$ inches. Experimenters may use width $2\frac{7}{16}''$ and length

Figure 68. Preparations to create one blue and one red triangle.

4″ for smaller rectangles, or width 4″ and length $6\frac{11}{16}''$ for larger rectangles. For each finished triangle, cut the shape in Figure 67 as follows. In the figure, the seam allowances are marked.

The darkened part is the equilateral triangle of side length d inches that will be visible in the end. Fold each rectangle in half along the dashed blue line. Cut off the left sides at a 30° angle from the tip of the fold, then cut off at 90° to create the top and bottom angles.

3. Each folded eventual-triangle shape (as in step 2) has two right angles. Align the one without folds with a corner of a square as in Figure 68. Lay another square on top, right sides together, and sew from the corner toward the acute angle; end this seam about $\frac{1}{4}''$ beyond the fold. This creates a triangle.

4. Open the sewn pieces, shape the triangle flaps, and pin the flaps down. Then turn over and trim the bulk from underneath the triangles as shown in Figure 69. We call these units *twosies*.

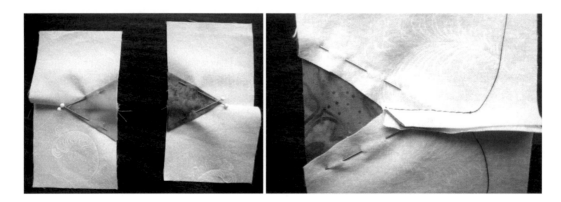

Figure 69. Open the blue and the red triangles, pin down the edges, and cut away from the seam allowances underneath. This forms two twosies.

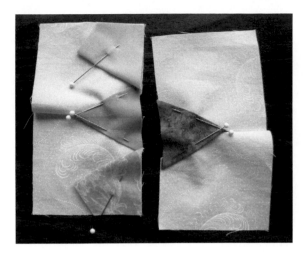

Figure 70. Preparing two twosies for seaming.

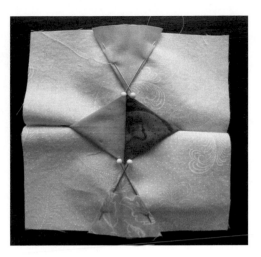

Figure 71. Two twosies sewn together into a foursie.

5. Prepare to add the green and yellow triangles as in Figure 70. Flip the right-hand unit over the left-hand unit and sew the right-hand seam. Now seam from edge to edge (not partially).

6. Open the sewn pieces, shape the two new triangle flaps and pin them down, then turn over and trim the bulk from underneath. You will obtain a foursie as in Figure 71.

7. Make as many foursies as needed. Merge them into eightsies as in Figure 72, and so on. Note that in merging foursies, a few stitches may have to be gently pulled apart near the blue and red vertices to make room for the yellow and green triangles.

The seams should be partial in areas where there is no sandwiched fabric, and from edge-to-edge otherwise. Always trim the bulk underneath the triangles before adding the next unit.

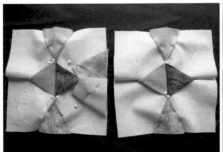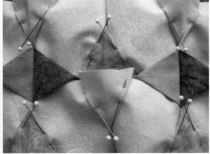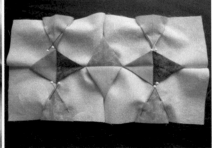

Figure 72. Two foursies sewn together into an eightsie.

Figure 73. Perhaps this intermediate step in the construction of the 6.6.6 tessellation can be developed further into a three-dimensional construction?

8. Press carefully at the end and trim to size, if necessary. Topstitch the loose triangle edges either now, or during the quilting process.

4.10   Construction of 6.6.6

This semiregular tessellation uses three colors. No two adjacent hexagons have the same color. Our construction fills a much-needed gap in the repertoire of quilting techniques. It is fussy, and uses excessive fabric and inordinate amounts of time. On the plus side, there is a very interesting three-dimensional intermediate configuration that might be worth pursuing in the future; see Figure 73.

Decide on the finished side length d of the hexagons. Draw out your design and determine how many hexagons of each color you will need. Say you need a hexagons of each fabric.

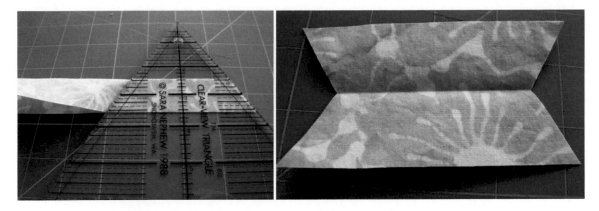

Figure 74. How to cut the eventual hexagon pieces from a strip (left); the desired eventual hexagon piece (right).

Materials

★ For each of the three hexagon fabrics, you need $\sqrt{3}d+$ 0.5 inches by a times $3d + \frac{\sqrt{3}}{2}$ inches.

Instructions

1. For hexagons with finished edge length d inches, cut fabric strips of width $\sqrt{3}d+0.5$ inches wide. Fold each strip in half lengthwise, without forming a crease. As in Figure 74, cut the folded strip at 60° angles to form an isosceles trapezoid whose parallel sides are along the edges of the strip, whose short parallel side is on the fold, and whose long edge is of length $3d+\frac{\sqrt{3}}{2}$ inches long. Experimenters may cut the fabric strips 3.5″ wide; the cut long edges should then be of length $6\frac{1}{16}''$.

   Cut a such fat-X shapes of each color; each of them will eventually produce one hexagon, hereafter referred to as an *eventual hexagon*.

2. We first sew the adjacent eventual hexagons for the *second* row; the setup is shown in Figure 75. Take any two of different colors, fold a pair of the third color lengthwise (as they were cut), and pin to the left-hand unfolded eventual hexagon.

Figure 75. How to prepare to sew *two* adjacent eventual hexagons.

Flip the right-hand eventual hexagon to make a sandwich and sew from edge to edge.

Repeat for all adjacent eventual hexagons in the second row, changing colors so that no two adjacent hexagons are of the same color. Figure 76 shows the outcome. You can see the hexagons forming in the middle (second) row. The top and the bottom rows have many loose edges.

Figure 76. This is *one* line of sewing, but uses eventual hexagons from *three* adjacent rows.

Figure 77. Merging two sets of three-row assemblies of hexagons.

3. Similarly, work on rows $4r + 1, 4r + 2, 4r + 3$, as r varies over the natural numbers, for as many rows as you planned.

4. The connecting of these units is analogous: as shown in Figure 77, place a row of eventual hexagons (with proper coloring) between two three-row assemblies of hexagons. Sew adjacent hexagons in this new row by inserting folded eventual hexagons from the row immediately above and the row immediately below, as in step 3.

Figure 78. There are still unsewn hexagon edges (left). Press carefully (right).

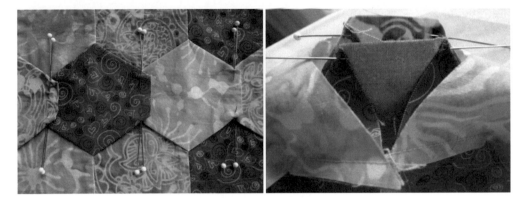

Figure 79. Pin to mark the desired seam lines (left), then pull to the other side to prepare for sewing (right).

Figure 80. The not-yet-quilted 6.6.6 tessellation.

5. Fewer edges are loose now. Press the formed creases carefully. See Figure 78.

6. In an ideal world, the unsewn edges would be precisely $\frac{1}{4}''$ from the desired seams. For the real world, mark the desired (or forced, as it may be) sewing lines with pins, then turn to the wrong side, and pull the pins and the unsewn edges through to the wrong side (see Figure 79). Sew along the marked lines.

   The result is a lot of hexagons, with many loose edges, as in Figure 80.

7. Press again, and topstitch the loose edges now or during the quilting stage.

Bibliography

[1] Chavey, Darrah P. "Periodic Tilings and Tilings by Regular Polygons, I: Bounds on the Number of Orbits of Vertices, Edges, and Tiles." *Mitteilungen Mathem. Seminar Giessen* 164 (1984), 37–50.

[2] Galebach, Brian. "n-Uniform Tilings." *Probability Sports*. Available at http://www.probabilitysports.com/tilings.html, 2010.

[3] Jahns, Marsha. *America's Heritage Quilts*, Better Homes and Gardens, Des Moines, IA, 1991.

[4] Grünbaum, Branko and Shephard, G. C. *Tilings and Patterns*. W. H. Freeman and Company, New York, 1987.

[5] Krötenheerdt, Otto. "Die Homogenen Mosaike *n*-ter Ordnung in der Euklidische Ebene," *Wiss. Z. Martin-Luther Univ. Halle-Wittenberg, Math.-Natur. Reihe* 18 (1969), 273–290.

[6] Wilson, Robin. *Four Colors Suffice*. Princeton University Press, Princeton, NJ, 2003.

ABOUT THE CONTRIBUTORS

The Editors are personally and professionally invested in connecting mathematics and fiber arts. sarah-marie focuses on knitting and has designed a number of objects, continually challenging herself to push the envelope of what is known and what is possible. Carolyn enjoys knitting, crocheting, tatting, and temari balls. Often she finds thinking deeply about the patterns of others to be as interesting as designing her own. Over the years we have made a concerted effort to foster thoughtful communities of fiber artists in various sectors of our lives. Most notably, we have held the Knitting Network at the AMS-MAA Joint Mathematics Meetings since 2000, at MathFest whenever we can, and at the AP Calculus readings in 2006 and 2009. We are also active members of local fiber arts groups.

Amy Szczepański has been knitting and crocheting since before she could read a pattern and picked up quilting and sewing after an art assignment in college relied on the visual language of American patchwork quilts. Classically trained as a mathematician with a PhD from the University of California at San Diego, she now works in education, outreach, and training at the Remote Data Analysis and Visualization Center at the University of Tennessee, where she helps people make sense of large datasets and learn computing tools for data visualization. She lives in Knoxville, Tennessee with her husband, Jim Conant, and two pet mice named Epsilon and Delta.

sarah-marie belcastro is a free-range mathematician whose primary mathematical research area is topological graph theory. She enjoys connecting people to each other, connecting ideas to each other, and connecting people to ideas. Among her many non-pure-mathematics interests are the feminist philosophy of science, dance (principally ballet and modern), and changing the world. sarah-marie did her undergraduate work at Haverford College and earned her PhD at the University of Michigan. For more, see her web domain at www.toroidalsnark.net, where her most popular pages are those on mathematical knitting.

Emily Peters is a mild-mannered mathematician by day, ravenous knitter by night. As a mathematician, she studies planar algebras. Planar algebras are a formalization of "proof by pictures" in algebra, and are related to operator algebras, knot theory, and field theories. As a knitter, she's recently become fascinated by the three-dimensional potential of knitting. In addition to cross-caps, she's made a garden of knobbly vegetables including a pepper, eggplant, beet, and carrot. Emily has a PhD from the University of California at Berkeley, where she was advised by Vaughan Jones. She currently holds an NSF postdoctoral fellowship at the Massachusetts Institute of Technology. She was taught to knit by sarah-marie.

Ted Ashton was introduced to tatting in high school. Before long he was mixing mathematics and tatting, and designed and created a tatted piece with three-way rotational symmetry. He tatted his way through undergraduate degrees in electrical engineering at Walla Walla University and mathematics at Southern Adventist University. He enjoys the creativity of designing his own patterns. The tatted Sierpiński triangle, created while earning his PhD at the University of Georgia, was his first foray into creating complex mathematical objects in fiber. Besides tatting, Ted enjoys playing chamber music, woodworking, and spending time with his wife Heidi (a knitter) and two children (who both enjoy numerous crafts).

Diane Herrmann teaches and is Co-Director of Undergraduate Studies in Mathematics at the University of Chicago. She is working toward Level II Certification in Canvaswork from the National Academy of Needlearts. She has won numerous awards for her work in NAN's Teacher Certification Program. She is a member of NAN, EGA, and ANG and has taught, served as program chair, and served as president for her local EGA chapter. Diane enjoys including multicultural ideas in her designs and finding ways to combine mathematics and the needlearts. She sings with Chicago's University Chorus and for several years sang with a barbershop quartet. She did her undergraduate work at Allegheny College and earned her PhD at the University of Chicago.

Mary Shepherd learned to knit at age six and began sewing doll clothes at age ten. She briefly taught various crafts at in-home sales parties akin to Tupperware parties. Her favorite of these crafts was (and still is) counted cross-stitch. She earned a Bachelor of Music from Missouri State University, a Masters in Accountancy from the University of Oklahoma and both an MA and a PhD in mathematics from Washington University in St. Louis. She teaches at Northwest Missouri State University and lives outside Savannah, Missouri on about four acres with her husband Ron, two dogs, and several cats.

Susan Goldstine enjoys making tactile and visual mathematical models, employing such diverse media as yarn, glass beads, copper tubes, pinecones, and pottery, though not all at once. She is also an avid cook, and while she usually pursues non-mathematical cookery, she hopes one day to reproduce the interlocking Escher swan cookies she made for a lark as an undergraduate. Susan received her BA in mathematics and French from Amherst College and her PhD in mathematics from Harvard University. She is currently an associate professor of mathematics at St. Mary's College of Maryland and finds that one of the best things about having tenure is being able to knit during faculty meetings.

Carolyn Yackel is an associate professor of mathematics at Mercer University specializing in mathematics and art, especially the fiber arts. She observes math of many genres arising in needlework and has developed a mathematics for liberal arts course centered around that theme. Carolyn has a panoply of divergent interests. For example, in addition to practicing a host of needlecrafts, she makes functional pottery, is an avid gardener, and loves to bake. Her undergraduate degree is from the University of Chicago and her PhD is from the University of Michigan.

Irena Swanson was introduced to quilting by Ellen Kirigin during her high school exchange year in Tooele, Utah. She started quilting more seriously during her third year in graduate school at Purdue University, where she earned her PhD. She is a professor of mathematics at Reed College, Portland, Oregon. Her office walls display several mathematics-related quilts.

CREDITS

Anything not credited has been created and/or photographed by the chapter author.

Chapter 1. The opening figure: models are Joshua, Jennifer, and Jonah Belden; crafting details are below. Figure 3 on page 18: students shown in sarah-marie's Spring 2010 Calculus II course at Mount Holyoke College are Cori Magnusson, Kim Finn, Faith Larson, Michelle Deveaux, Jordan Seto, Alex Bemis, Alissa Correll, Christina Horimoto, Sophia Weeks, and May Aunaetitrakul. Figure 6 on page 20: hat modeled by Jonah Belden, knitted by Zia Marek-Loftus in Lambs Pride worsted (Onyx) and Cascade 220 (goldenrod). I-cord stinger (co 5 st, knit 8 rounds, k2tog twice, k3tog) by sarah-marie belcastro. Figure 7 on page 22: hat modeled by Sean Kinlin, knitted by sarah-marie belcastro in yarn handspun by Karen Robinson. Figure 8 on page 24: left hat modeled by Rachel Shorey and knitted by Angela Juliet Larke using Berroco Comfort, one skein each of colors 9710, 9726, and 9737; middle hat modeled by Jennifer Belden and knitted by Carolyn Yackel using Valley yarns Sheffield, one skein each of Teal (07), Navy (10), and Brick Red (14); right hat modeled by Chelsea Land and knitted by Amy Szczepański in Cascade Pastaza, colors 6002, 013, and 010. Diagrams created by Amy using Lineform and by Carolyn using Illustrator. Photographs by sarah-marie belcastro.

Chapter 2. The opening figure and Figure 18 on page 43: nightcap and socks modeled by Tom Hull and knitted in Di.Ve' Zenith in colors turquoise, violet, blue ice, and grape with cornflower sock heels/toes. Figure 1 on page 29: sample knitted in Paas-dyed wool and suspected Cascade 220. Figure 2 on page 30: the UMass Amherst CS Department Information Extraction and Synthesis Laboratory's tea cozy was designed and knit by Rachel Shorey using Bernat Berella 4 Solids & Ragg colorways Rich Periwinkle Blue and Natural. Figure 19 on page 43: upper socks modeled by Tom Hull; lower socks modeled by Jennifer Belden and knitted by Sharon Frechette. Figure 20 on page 46: hat knitted by Sharon Frechette in Berroco Vintage, colors Dewberry (5167), Wasabi (5165), Cast Iron (5145), and Wild Blueberry (5160). Figure 21 on page 47: socks modeled by Tom Hull. Diagrams created by sarah-marie using Canvas and Carolyn using Illustrator.

Chapter 3. The opening figure: projective plane knitted by Carolyn Yackel in Knit One, Crochet Too Paintbox colorway Adobe Rose (02), surrounded by wooden dissection puzzles and found bird's nest. Photograph by sarah-marie belcastro. Puzzles, left to right, are Star of David designed and crafted by Stewart Coffin, Dual Tetrahedron designed and crafted by Wayne Daniel, Domino Tower designed by Oskar van Deventer and crafted by Scott Petersen, 32 Corner Ball designed and produced by Josef Pelikan, Sliparoo designed and crafted by Stewart Coffin, Iso-Prism designed and crafted by Stewart Coffin, D–H–T designed and crafted by Wayne Daniel, Octahedrik designed by Rik Brouwer and produced by Josef Pelikan, and Pennyhedron variant designed and crafted by Stewart Coffin. Figure 1 on page 51: created by Carolyn Yackel in Adobe Illustrator. Figures 2, 7, 8, and 9 on pages 52, 56, 56, and 57: cross-cap and partial cross-caps knitted in Noro Kureyon colorway 40U. Diagrams created using Tikz.

Chapter 4. The opening figure: from left to right, beading by Miriam Roberts, cross-stitch by Jennifer Belden, tatting by John Nance, cross-stitch by Jennifer Belden, beading by Katharine Merow, cross-stitch by Diane Herrmann, tatting by Carolyn Yackel, and beading by Miriam Roberts. Figure 27 on page 74: tatting by Carolyn Yackel in Circulo

Dalila Egyptian cotton thread; rubber monsters gift of Joan and Frank Belcastro (they light up in multiple colors!); photograph by sarah-marie belcastro. Figure 30 on page 76: tatting by John Nance in H. H. Lizbeth thread, size 20, color 651, using a Boyle shuttle; photograph by sarah-marie belcastro. Figure 31 on page 77: upper Triangle beaded by Katharine Merow; middle and lower Triangles beaded by Miriam Roberts. Figure 32 on page 78: beading by Miriam Roberts. Figure 35 on page 79: string art by Dan Zook in craft thread, photographs by Carolyn Yackel. Figure 38 on page 81: top triangle cross-stitched by Jennifer Belden, lower two triangles cross-stitched by Diane Herrmann; photograph by sarah-marie belcastro. Eggs graciously supplied by the fowl pictured in Figure 6 on page 20, who live next door to sarah-marie. Figure 41 on page 83: left triangle cross-stitched by Jennifer Belden, right triangle cross-stitched by Diane Herrmann; photograph by sarah-marie belcastro. Figures 43 and 47 on pages 83 and 85: triangles cross-stitched by Diane Herrmann; photograph by sarah-marie belcastro. Figure 45 on page 84: triangle cross-stitched by Jennifer Belden; photograph by sarah-marie belcastro. Remaining photos by Heidi Ashton. Diagrams created using TikZ.

Chapter 5. The opening figure: left cube stitched by Mary Shepherd in knitting yarn on plastic canvas; middle cube stitched by Diane Herrmann and right cube stitched by Kelly Meyer, both in Paternayan® yarn on mono canvas; photograph by sarah-marie belcastro. Figure 1 on page 89: cushion stitched in Paternayan® yarn by Vilma Herrmann; designer unknown. Figure 2 on page 90: detail of *Your Roots are Showing!*, Diane Herrmann © 2009. Figure 8 on page 95: detail of area stitched with Silk 'n Ivory wool from *brown paper packages*. Figures 9, 16, and 17 on pages 95, 98, and 99: photos by sarah-marie, stitching by Diane in Paternayan® yarn. Figure 15 on page 97: photo by sarah-marie; stitching by Diane in Rainbow Gallery threads, including Neon Rays and Fyre Works. Figure 23 on page 103: photo by sarah-marie; stitching by Diane in Paternayan® yarn. Figure 25 on page 105: photo by sarah-marie; stitching by Lori Schreiber in DMC pearl cotton #5 on 18 count mono canvas. Diagrams created by Carolyn using Illustrator.

Chapter 6. The opening figure: photograph by sarah-marie belcastro. Figure 26 on page 132: stitching by Hope McIlwain; photograph by Carolyn Yackel. Diagrams created by Carolyn using Illustrator. Photos by Neil Hatfield and photo editing by sarah-marie belcastro.

Chapter 7. The opening figure: pastel blankets crocheted by Ida Thornton in acrylic yarn; brightly colored blanket crocheted by Carolyn Yackel in cotton. Model is Jonah Belden; photograph by sarah-marie belcastro. Figure 3 on page 143: hanging crocheted in Debbie Bliss baby cashmerino. Figure 7 on page 147: baby blanket crocheted by Carolyn Yackel in Gedifra Mayra pink (2057), pale green (2067), and citron (2028); photography by Carolyn. Diagrams created by Susan and edited by Carolyn using Illustrator.

Chapter 8. Figures 20, 22, 24, 26, 28, and 30 on pages 163–166 were graciously supplied by Chaim Goodman-Strauss and are reprinted from *The Symmetry of Things* (A K Peters, 2008). Background removal on Figures 8, 9, 21, 23, 25, 27, 29, and 31 on pages 156, 156, and 163–166 by Thomas Hull using Photoshop. Figures 15, 16, and 17 on pages 160, 161, and 162 by sarah-marie belcastro using *Mathematica*. Figure 35 on page 171: left-hand ball by Leslie Hodges, middle ball and photo by Susan Schmoyer, and right-hand ball by Margaret Eilrich with photo by Mark Eilrich. Figure 43 on page 176: middle ball by Leslie Hodges and right-hand ball by Margaret Eilrich with photo by Mark Eilrich. Figure 50 on page 180: second ball and photo by Susan Schmoyer, third ball by Margaret Eilrich and photo by Mark Eilrich, and last ball by Leslie Hodges. Diagrams created using Illustrator. Photo editing by sarah-marie belcastro.

Chapter 9. All quilts created using 100% cotton fabrics. The opening figure: photograph by Marko Šifrar. Figure 15 on page 198: sampler quilt by Jeri Riggs using cotton batik fabrics, cotton batting, and rayon thread; photo by sarah-marie belcastro. Jeri Riggs studied mathematics and psychology at MIT and Wellesley College and earned her M.D. from Boston University. Since retiring from practicing psychiatry, she has enjoyed making quilts and designing knitwear that incorporates her love of color, geometry, and line. She has exhibited in quilt shows nationally, including the American Quilter's Society Shows and the International Quilt Festival. You can see more of her quilts at http://www.jeririggs.com and follow her knitwear work in progress at http://jeririggs.blogspot.com. Figure 17 on page 202: table runner by Joan Belcastro; photo by sarah-marie belcastro. Figure 30 on page 207: Veronika Šifrar standing on quilt; photo by Marko Šifrar. Diagrams created by Irena using Macaulay 2 (exported to pstricks)

and sarah-marie belcastro using Canvas. Photo editing by sarah-marie.

About the Contributors. sarah-marie belcastro's headshot by Thomas C. Hull. Emily Peters's headshot by Kelly Michaud. Ted Ashton's headshot by his wife Heidi Ashton. Diane Herrmann's headshot by Laurie Wall. Mary D. Shepherd's headshot by Neil Hatfield. Susan Goldstine's headshot by Marian Goldstine. Susan's wrap is the Koigu Alligator Wrap by Maie Landra in Koigu Painter's Palette Premium Merino (KPPPM), colors P836, P148, P431, P621, and P603. Carolyn Yackel's headshot by Gerald R. Lucas. Irena Swanson's headshot by her husband Steve Swanson.

Other opening figures. Opening figure on p. vi: cube stitched by Diane Herrmann, snake by Nature; photo by sarah-marie belcastro. Opening figure on p. x: socks modeled by Jenn Belden and Tom Hull; blue socks knitted by sarah-marie and pink socks knitted by Sharon Frechette; photo by sarah-marie. Opening figure on p. 0: model is Lolita Rosental; blanket crocheted by Carolyn Yackel; photo by Wyatt Rosental. Opening figure on p. 238: closeup of 4.4.4.4 section of sampler quilt by Jeri Riggs; photo by sarah-marie. Opening figure on p. 242: closeup of 3.3.4.3.4 section of sampler quilt by Jeri Riggs; photo by sarah-marie. Opening figure on p. 246: closeup of 3.12.12 section of sampler quilt by Jeri Riggs; photo by sarah-marie.

FIBER ARTS BIBLIOGRAPHY

All Fiber Arts Except Weaving

[1] Adams, Colin, Fleming, Thomas, and Koegel, Christopher. "Brunnian Clothes on the Runway: Not for the Bashful." *American Mathematical Monthly* 111:9 (2004), 741–748.

[2] belcastro, sarah-marie. "Every Topological Surface Can Be Knit: A Proof." *Journal of Mathematics and the Arts* 3:2 (2009), 67–83.

[3] belcastro, sarah-marie and Yackel, Carolyn. "About Knitting." *Math Horizons* (November 2006), 24–27, 39.

[4] belcastro, sarah-marie and Yackel, Carolyn, eds. *Making Mathematics with Needlework*. A K Peters, Wellesley, MA, 2008.

[5] Bernasconi, Anna, Bodei, Chiara, and Pagli, Linda. "Knitting for Fun: A Recursive Sweater." In *Fun with Algorithms*, Lecture Notes in Computer Science 4475, edited by Pierluigi Crescenz, Giuseppe Prencipe, and Geppino Pucci, pp. 53–65. Springer, New York, 2007.

[6] Bernasconi, Anna, Bodei, Chiara, and Pagli, Linda. "On Formal Descriptions for Knitting Recursive Patterns." *Journal of Mathematics and the Arts*, 2:1 (March 2008), 9–27.

[7] Biedl, Therese, Horton, John D., and Lopez-Ortiz, Alejandro, "Cross-Stitching Using Little Thread." In *Proceedings of the 17th Canadian Conference on Computational Geometry (CCCG'05)*, pp. 199–202. Available at http://www.cccg.ca/proceedings/2005/54.pdf.

[8] Cochrane, Paul. "Knitting Maths." *Mathematics Teaching* (September 1988), 26–28.

[9] Conway, J. H. "Mrs. Perkins's Quilt." *Mathematical Proceedings of the Cambridge Philosophical Society* 60 (1964), 363–368.

[10] Curtis, S. A. "An Application of Functional Programming: Quilting." In *Trends in Functional Programming*, Vol. 2 (Proceedings of the 2nd Scottish Functional Programming Workshop, St. Andrew's, 2000), edited by Stephen Gilmore, pp. 145–158. Intellect Books, Exeter, UK, 2000.

[11] Curtis, S. A. "Marble Mingling." *Journal of Functional Programming*, 16:2 (2006), 129–136.

[12] DeTemple, Duane. "Reflection Borders for Patchwork Quilts." *Mathematics Teacher* (February 1986), 138–143.

[13] Ellison, Elaine Krajenke. "Imaginative Quilted Geometric Assemblages." In *Bridges Donostia: Proceedings 2007*, pp. 425–426. Tarquin Publications, London, 2007.

[14] Fisher, Gwen. "Quilt Designs Using Non-Edge-to-Edge Tilings by Squares." In *Meeting Alhambra, ISAMA-BRIDGES Conference Proceedings 2003*, edited by R. Sarhangi and C. Sequin, pp. 265–272. University of Granada Publication, Granada, 2003.

[15] Fisher, Gwen. "The Quaternions Quilts." *FOCUS* 25:1 (2005), 4–5.

[16] Fisher, Gwen and Medina, Elsa. "Cayley Tables as Quilt Designs." In *Meeting Alhambra, ISAMA-BRIDGES Conference Proceedings 2003*, edited by R. Sarhangi and C. Sequin, pp. 553–554. University of Granada Publication, Granada, 2003.

[17] Funahashi, Tatsushi, Yamada, Masashi, Seki, Hirohisa, and Itoh, Hidenori. "A Technique for Representing Cloth Shapes and Generating 3-Dimensional Knitting Shapes." *Forma* 14:3 (1999), 239–248.

[18] Griffin, Mary. "Wear Your Own Theory!: A Beginner's Guide to Random Knitting." *New Scientist* (March 26, 1987), 69–70.

[19] Grishanov, S. A., Cassidy, T, and Spencer, D. J. "A Model of the Loop Formation Process on Knitting Machines Using Finite Automata Theory." *Applied Mathematical Modelling* 21:7 (July 1997), 455–465.

[20] Grünbaum, Branko. "Periodic Ornamentation of the Fabric Plane: Lessons from Peruvian Fabrics." In *Symmetry Comes of Age: The Role of Pattern in Culture*, edited by Dorothy K. Washburn and Donald W. Crowe, pp. 18–64, University of Washington Press, Seattle, 2004.

[21] Harris, Mary. "Mathematics and Fabrics." *Mathematics Teaching* 120 (1987), 43–45.

[22] Harris, Mary." *Common Threads: Women, Mathematics and Work*. Trentham Books, Stoke-on-Trent, 1997.

[23] Henderson, David W. and Taimina, Daina. "Crocheting the Hyperbolic Plane." *The Mathematical Intelligencer* 23:2 (2001), 17–28.

[24] Igarashi, Yuki, Igarashi, Takeo, and Suzuki, Hiromasa. "Knitting a 3D Model." *Computer Graphics Forum (Proceedings of Pacific Graphics 2008)* 27:7 (Oct. 2008), 1737–1743.

[25] Irving, Claire. "Making the Real Projective Plane." *Mathematical Gazette* (November 2005), 417–423.

[26] Isaksen, Daniel and Petrofsky, Al. "Mobius Knitting." In *Bridges: Mathematical Connections in Art, Music, and Science*, edited by R. Sarhangi, pp. 67–76. Southwestern College, Winfield, KS, 1999.

[27] Kaldor, Jonathan M., James, Doug L., and Marschner, Steve. "Simulating Knitted Cloth at the Yarn Level." *International Conference on Computer Graphics and Interactive Techniques, ACM SIGGRAPH 2008 Papers*, Article No. 65. ACM Press, New York, 2008.

[28] Mabbs, Louise. "Fabric Sculpture—Jacob's Ladder." In *Bridges London: Conference Proceedings 2006*, edited by R. Sarhangi and J. Sharp, pp. 561–568. Tarquin Publications, London, 2006.

[29] Mallos, James. "Triangle-Strip Knitting." *Hyperseeing: Proceedings of ISAMA 2010* (Summer 2010), 111–116.

[30] Osinga, Hinke M. and Krauskopf, Bernd. "Crocheting the Lorenz Manifold." *The Mathematical Intelligencer* 26:4 (2004), 25–37.

[31] Osinga, Hinke M. and Krauskopf, Bernd. "The Lorenz Manifold: Crochet and Curvature." In *Bridges London: Conference Proceedings 2006*, edited by R. Sarhangi and J. Sharp, pp. 255–260. Tarquin Publications, London, 2006.

[32] Osinga, Hinke M. and Krauskopf, Bernd. "Visualizing Curvature on the Lorenz Manifold." *Journal of Mathematics and the Arts* 1:2 (2007), 113–123.

[33] Pickett, Barbara Setsu. "Sashiko: the Stitched Geometry of Rural Japan." In *Bridges London: Conference Proceedings 2006*, edited by R. Sarhangi and J. Sharp, pp. 211–214. Tarquin Publications, London, 2006.

[34] Reid, Miles. "The Knitting of Surfaces." *Eureka—The Journal of the Archimedeans* 34 (1971), 21–26.

[35] Ross, Joan. "How to Make a Möbius Hat by Crocheting." *Mathematics Teacher* 78 (1985), 268–269.

[36] Shepherd, Mary D. "Groups, Symmetry and Other Explorations with Cross Stitch." *Electronic Proceedings of the Missouri MAA*. Available at http://www.missouriwestern.edu/orgs/momaa/, 2007.

[37] Trustrum, G. B. "Mrs Perkins's Quilt." *Mathematical Proceedings of the Cambridge Philosophical Society* 61 (1965), 7–11.

[38] Williams, Mary C. "Quilts Inspired by Mathematics." In *Meeting Alhambra, ISAMA-BRIDGES Conference Proceedings 2003*, edited by R. Sarhangi and C. Sequin, pp. 393–399. University of Granada Publication, Granada, 2003.

[39] Williams, Mary C. and Sharp, John. "A Collaborative Parabolic Quilt." In *Bridges: Mathematical Connections in Art, Music, and Science, Conference Proceedings 2002*, edited by R. Sarhangi, pp. 143–149. Southwestern College, Winfield, KS, 2002.

[40] Woolfit, Penelope. "The Geometry of Asian Trousers." In *Bridges Donostia: Proceedings 2007*, pp. 427–430. Tarquin Publications, London, 2007.

[41] Yackel, C. A. "Embroidering Polyhedra on Temari Balls." In *Math+Art=X Conference Proceedings*, pp. 183–187. University of Colorado, Boulder, CO, 2005.

[42] Yackel, C. A. "Marking a Physical Sphere with a Projected Platonic Solid." In *Bridges Banff: Proceedings 2009*, pp. 123–130. Tarquin Publications, London, 2009.

Weaving

[43] Akleman, Ergun, Chen, Jianer, Xing, Qing, and Gross, Jonathan L. "Cyclic Plain-Weaving on Polygonal Mesh Surfaces with Graph Rotation Systems." *ACM Transactions on Graphics: Proceedings of ACM SIGGRAPH 2009* 28:3 (August 2009), Article No. 78.

[44] Chen, Yen-Lin, Akleman, Ergun, Chen, Jianer, and Xing, Qing. "Designing Biaxial Textile Weaving Patterns." *Hyperseeing: Proceedings of ISAMA 2010* (Summer 2010), 53–62.

[45] Clapham, C. R. J. "When a Fabric Hangs Together." *Bulletin of the London Mathematical Society* 12:3 (1980), 161–164.

[46] Clapham, C. R. J. "The Bipartite Tournament Associated with a Fabric." *Discrete Mathematics* 57:1–2 (1985), 195–197.

[47] Clapham, C. R. J. "When a Three-Way Fabric Hangs Together." *Journal of Combinatorial Theory Series B* 38:2 (1985), 190.

[48] Clapham, C. R. J. "The Strength of a Fabric." *Bulletin of the London Mathematical Society* 26:2 (1994), 127–131.

[49] Delaney, Cathy. "When a Fabric Hangs Together." *Ars Combinatoria* 21-A (1986), 71–79.

[50] Enns, T. C. "An Efficient Algorithm Determining When a Fabric Hangs Together." *Geometriae Dedicata*, 15 (1984) 259–260.

[51] Grünbaum, B. and Shephard, G. C. "Satins and Twills: Introduction to the Geometry of Fabrics." *Mathematics Magazine* 53:3 (1980), 139–161.

[52] Grünbaum, B. and Shephard, G. C. "A Catalogue of Isonemal Fabrics." *Discrete Geometry and Convexity, Annals of the New York Academy of Sciences* 440 (1985), 279–298.

[53] Grünbaum, B. and Shephard, G. C. "An Extension to the Catalogue of Isonemal Fabrics." *Discrete Mathematics* 60 (1986), 155–192.

[54] Grünbaum, B. and Shephard, G. C. "Isonemal Fabrics." *American Mathematical Monthly* 95 (1988), 5–30.

[55] Hoskins, J. A. "Factoring Binary Matrices: A Weaver's Approach." In *Combinatorial Mathematics, IX (Brisbane, 1981)*, Lecture Notes in Mathematics 952, pp. 300–326. Springer, Berlin-New York, 1982.

[56] Hoskins, J. A. "Binary Interlacement Arrays and Structural Cross-Sections." *Congressus Numerantium* 40 (1983), 63–76.

[57] Hoskins, Janet A. and Hoskins, W. D. "The Solution of Certain Matrix Equations Arising from the Structural Analysis of Woven Fabrics." *Ars Combinatoria* 11 (1981), 51–59.

[58] Hoskins, J. A., Hoskins, W. D., Street, Anne Penfold, and Stanton, R. G. "Some Elementary Isonemal Binary Matrices." *Ars Combinatoria* 13 (1982), 3–38.

[59] Hoskins, J. A. and Hoskins, W. D. "An Algorithm for Color Factoring a Matrix." In *Current Trends in Matrix Theory (Auburn, AL, 1986)*, pp. 147–154. North-Holland, New York, 1987.

[60] Hoskins, Janet A., Praeger, Cheryl E., Street, and Anne Penfold. "Balanced Twills with Bounded Float Length." *Congressus Numerantium* 40 (1983), 77–89.

[61] Hoskins, Janet A., Praeger, Cheryl E., Street, and Anne Penfold. "Twills with Bounded Float Length." *Bulletin of the Australian Mathematical Society* 28:2 (1983), 255–281.

[62] Hoskins, J. A., Stanton, R. G., and Street, Anne Penfold. "Enumerating the Compound Twillins." *Congressus Numerantium* 38 (1983), 3–22.

[63] Hoskins, J. A., Stanton, R. G., and Street, A. P. "The Compound Twillins: Reflection at an Element." *Ars Combinatoria* 17 (1984), 177–190.

[64] Hoskins, Janet A., Street, Anne Penfold, and Stanton, R. G. "Binary Interlacement Arrays, and How to Find Them." *Congressus Numerantium* 42 (1984), 321–376.

[65] Hoskins, J. A., Thomas, R. S. D. The Patterns of the Isonemal "Two-Colour Two-Way Two-Fold Fabrics." *Bulletin of the Australian Mathematical Society* 44:1 (1991), 33–43.

[66] Lucas, E. *Application de l'Arithmétique à la Construction de l'Armure des Satins Réguliers*. Retaux, Paris, 1867.

[67] Lucas, E. "Principii fondamentali della geometria dei tessute." *L'Ingegneria Civile e le Arti Industriali* 6 (1880) 104–111, 113–115.

[68] Lucas, E. "Les principes fondamentaux de la géometrie des tissus," *Compte Rendu de L'Association Française fpour l'Avancement des Sciences* 40 (1911), 72–88.

[69] Oates-Williams, Sheila, and Street, Anne Penfold. "Universal Fabrics." In *Combinatorial Mathematics VIII: Proceedings of the Eighth Australian Conference on Combinatorial Mathematics Held at Deakin University, Geelong, Australia, August 25–29, 1980*, Lecture Notes in Mathematics 884, pp. 355–359. Springer, Berlin, 1981.

[70] Pedersen, Jean J. "Some Isonemal Fabrics on Polyhedral Surfaces." In *The Geometric Vein: The Coxeter Festschrift*, edited by Chandler Davis, Branko Grünbaum, and F. A. Sherk, pp. 99–122. Springer-Verlag, New York, 1981.

[71] Pedersen, Jean. "Geometry: The Unity of Theory and Practice." *The Mathematical Intelligencer* 5:4 (1983), 37–49.

[72] Philips, Tony. "Inside-Out Frieze Symmetries in Ancient Peruvian Weavings." *AMS Feature Column*. Available at http://www.ams.org/featurecolumn/archive/weaving.html, October 2008.

[73] Roth, Richard L. "The Symmetry Groups of Periodic Isonemal Fabrics." *Geometriae Dedicata* 48 (1993), 191–210.

[74] Roth, Richard L. "Perfect Colorings of Isonemal Fabrics Using Two Colors." *Geometriae Dedicata* 56 (1995), 307–326.

[75] Shorter, S. A. "The Mathematical Theory of the Sateen Arrangement." *The Mathematical Gazette* 10 (1920), 92–97.

[76] Thomas, R. S. D. "Isonemal Prefabrics with Only Parallel Axes of Symmetry." *Discrete Mathematics* 309:9 (6 May 2009), 2696–2711.

[77] Thomas, R. S. D. "Isonemal Prefabrics with Perpendicular Axes of Symmetry." *Utilitas Mathematica* 82 (2010), 33–70.

[78] Thomas, R. S. D. "Isonemal Prefabrics with No Axes of Symmetry." *Discrete Mathematics* 310:8 (2010), 1307–1324.

[79] Thomas, R. S. D. "Perfect Colourings of Isonemal Fabrics by Thin Striping." *Bulletin of the Australian Mathematical Society*, to appear.

[80] Woods, H. J. "Part II—Nets and Sateens, of The Geometrical Basis of Pattern Design." *Textile Institute of Manchester Journal* 26 (1935), T293–T308.

[81] Zelinka, Bohdan. "Isonemality and Mononemality of Woven Fabrics." *Applications of Mathematics* 28:3 (1983), 194–198.

[82] Zelinka, Bohdan. "Symmetries of Woven Fabrics." *Applications of Mathematics* 29:1 (1984), 14–22.

INDEX

Crafting by Concepts

Designed by Erica Schultz
   Typeset by A K Peters, Ltd.
   Printed and bound by Replika Press, Pvt, India

Composed in Adobe Myriad using LaTeX
   Printed on 90 gsm Gloss Art Paper

T - #0770 - 101024 - C264 - 229/204/12 [14] - CB - 9781568814353 - Matt Lamination